INTERPRETING OLD IRONSIDES

Making Old Ironsides' storied past come alive, Seaman Stephen Bland addresses his tour group on the frigate's spar deck. *Photo by MC1 Charles L. Ludwig, USN*

Charles E. Brodine, Jr.
Michael J. Crawford
Christine F. Hughes

NAVAL HISTORICAL CENTER
DEPARTMENT OF THE NAVY
WASHINGTON 2007

Secretary of the Navy's Advisory Subcommittee on Naval History

Published by
Naval Historical Center
805 Kidder Breese Street SE
Washington Navy Yard, DC 20374-5060
www.history.navy.mil
Contact us at 202-433-9785

Book design by Marco Marchegiani

Front cover: USS *Constitution* Escaping the British, July 1812. *Julian O. Davidson, oil on canvas, 1884. Reproduced courtesy of the USS Constitution Museum, Boston, Massachusetts*
Back cover: Heritage. *Naval Historical Center photograph*

U.S. GOVERNMENT OFFICIAL EDITION NOTICE

Use of ISBN
This is the Official U.S. Government edition of this publication and is herein identified to certify its authenticity. Use of the 978-0-945274-54-4 is for U.S. Government Printing Office Official Editions only. The Superintendent of Documents of the U.S. Government Printing Office requests that any reprinted edition clearly be labeled as a copy of the authentic work with a new ISBN.

Library of Congress Cataloging-in-Publication Data

Brodine, Charles E., 1956-
Interpreting Old Ironsides : an illustrated guide to USS Constitution / Charles E. Brodine, Jr., Michael J. Crawford, Christine F. Hughes.
p. cm.
Summary: "Collection of richly illustrated essays and transcribed original documents highlighting the history of America's oldest commissioned ship, USS Constitution, and naval life in the age of sail" —Provided by publisher.
Includes bibliographical references and index.
ISBN 978-0-945274-54-4 (alk. paper)
1. Constitution (Frigate). 2. Constitution (Frigate)—Pictorial works. 3. United States. Navy—Sea life—History—19th century. I. Crawford, Michael J. II. Hughes, Christine F., 1949- III. Title.
VA65.C7B76 2007
359.3'220973—dc22

2007020836

♾ The paper used in this publication meets the requirements for permanence established by the American National Standard for Information Sciences "Permanence of Paper for Printed Library Materials" (ANSI Z39.48-1984).

For sale by the Superintendent of Documents, U.S. Government Printing Office
Internet: Bookstore.gpo.gov
Phone: toll free 866-512-1800; DC area 202-512-1800
Fax: 202-512-2104
Mail: Stop SSOP, Washington, DC 20402-0001

ISBN 978-0-945274-54-4 (alk. paper)

·ontents

IV. Appendices

Foreword

At the close of the War of 1812, in recognition of the United States frigate *Constitution's* remarkable victories during that conflict, the editor of the *National Intelligencer* proposed formally recognizing the revered warship's status as a national icon. "Let us keep 'Old Iron Sides' at home," he urged. "She has, literally become a Nation's Ship, and should be preserved. Not as a 'sheer hulk, in ordinary' (for she is no ordinary vessel); but, in honorary pomp, as a glorious Monument of her own, and our other National Victories."

Despite this call to preserve *Constitution* as a permanent, floating memorial to the Navy's wartime achievements, the Navy continued to employ *Constitution* on active service through much of the nineteenth century. The 1897 centennial of its launching inaugurated a series of calls to restore the venerable but now deteriorated ship and preserve it as a permanent museum. Over the course of the twentieth century, in response to the voice of the people of the United States, who have taken Old Ironsides to their hearts as a symbol of national valor, Congress provided for the ship's restoration and preservation.

The current operative law embodying the public will to preserve *Constitution* as a national icon dates to July 23, 1954. Public Law 523, 83d Congress, authorizes the Secretary of the Navy "to repair, equip, and restore the United States Ship *Constitution* as far as may be practicable, to her original condition, but not for active service, and thereafter to maintain the United States Ship *Constitution* at Boston, Massachusetts." The Naval Historical Center takes pride in the responsibility entrusted to it under this act to maintain and interpret Old Ironsides with historical authenticity.

Interpreting Old Ironsides is a valuable tool for use in fulfilling this trust. This book is a training guide for the seamen stationed in the ship and given the responsibility as tour guides of explaining to the public its important place in the nation's history. Few of these Sailors, 60 percent of whom arrive straight from boot camp, begin with more than a rudimentary knowledge of *Constitution*'s storied past and all go through a process of thorough training. Tour guides receive instruction both in the ship's history and in the techniques of tour guiding. By helping the seamen master the history, this guide bolsters the confidence with which they interact face-to-face with the public.

The essays in *Interpreting Old Ironsides* will appeal to anyone, not just the Sailors assigned to the ship, curious about the history of the frigate *Constitution*. While the Sailors take just pride in their ship, so may all citizens, for, as the editor of the *National Intelligencer* proclaimed nearly two hundred years ago, *Constitution* is the "Nation's Ship."

RADM Paul E. Tobin, Jr., USN (Ret.)
Director, Naval Historical Center

Preface

Around noon, on the 21st of October 1797, the 44-gun frigate *Constitution* slid down the ways at Hartt's shipyard into the waters of Boston Harbor. Neither the day's cold temperatures nor its overcast skies dampened the spirits of those in attendance as they cheered the launch of the powerful new American warship into its natural element. As heartening as *Constitution*'s launch was to Boston's proud townspeople, as a symbol of local ingenuity and of national resolve to protect the country's oceangoing commerce, surely no one who viewed the launch that October day imagined that over two hundred years later *Constitution* would still be afloat serving the Navy.

Yet afloat *Constitution* remains, having survived three wars, service on numerous distant stations, duty as a training and receiving ship, extended periods of neglect and decay, and occasional candidacy for the ship-breaker's yard. That the distinguished frigate has endured all these tribulations to become the oldest commissioned warship afloat marks *Constitution* as, in the words of one of its modern biographers, "a most fortunate ship."

Today *Constitution* is the ceremonial flagship of the fleet, crewed by active duty officers and Sailors. It is used for public and military ceremonies, the most noteworthy of which are the ship's annual turnaround cruises. But *Constitution*'s most important mission is educational, and its crew, through guided tours and school outreach programs such as "Old Ironsides across the Nation," helps foster a greater public awareness of America's rich naval history, customs, and traditions.

This book arose out of a need to provide the men and women now serving in *Constitution* with an authoritative resource to assist them in interpreting the ship to the public. With the guidance of *Constitution*'s ship's historian, the staff of the Naval Historical Center's Early History branch prepared and assembled lesson plans, documents, and illustrations that chronicle the history of *Constitution*. The lesson plans were drafted to provide succinct summaries on various topics or themes relating to the ship. Although they may be read in any order, they have been grouped in such a way as to establish a core knowledge of the ship that can be built upon as the reader progresses through the book, with later sections adding greater depth of knowledge and fuller historical context of the ship and its times.

The majority of the essays focus on the ship's operational days, especially its service in the War of 1812—the period of which the frigate is portrayed to the public today. Other articles enlarge on *Constitution*'s postwar career and describe the world of the serving Sailor in the age of sail. Appendices accompanying the text enhance and amplify many of the lesson plans. For example, for each one of *Constitution*'s 1812 actions the reader will be able to consult not only a brief essay describing the action itself but also the official after-action reports of each captain. Likewise, documents are included that reveal how the men of the ship dressed, ate, and were disciplined. Numerous illustrations depict the ship and its crew in battle and at work at various stages in its career. This book seeks then, not only to tell *Constitution*'s story but also to capture the experience of the serving Sailor in the early sailing Navy.

Though this guide was originally conceived for a very small and select audience (*Constitution*'s crew), the range of topics it covers, the wealth of illustrative and documentary material that accompanies it, and the style in which the essays have been written, make it worth-

while reading for anyone intrigued by Old Ironsides and the early American Navy. History enthusiasts of many stripes—naval, military, maritime, social, diplomatic—will also find much to engage their interest between the covers of this book.

It should be noted that while these lesson plans provide detailed information on *Constitution* and its times, they are not intended to provide a comprehensive history of the ship's career. Interested readers will find within, a selected bibliography to consult for additional books, articles, and websites to broaden their knowledge of this historic vessel.

The events related in this book should remind the reader that the role of the Navy has changed little from the day when *Constitution* was launched—that is, to defend and advance the nation's interests in home and foreign waters. This book is offered to the public to promote and celebrate the history of a remarkable ship and the men and women who have served in it. It is hoped in turn that this will deepen the American public's appreciation of its maritime heritage in general and the part that the U.S. Navy played in creating that heritage in particular.

From its inception, this project has received the enthusiastic support of *Constitution*'s commanders and crew. In particular, BM1(SW) Andrew P. Dingman, FCC(SW) Andrew P. Wenzel, and EM1(SW) Aaron M. Walker, who, serving successively as ship's historian while this project was underway, gave helpful guidance on the book's content and organization. The authors also benefited from the suggestions and comments of Anne Rand, Sarah Watkins, and Kristin Gallas of USS Constitution Museum who reviewed an early draft of this manuscript. Kate Lennon Walker and Harrie Slootbeek, also of USS Constitution Museum, responded promptly to our numerous requests for scanned images from that institution's manuscript and art collections. As with all publications undertaken by our office, we have received the generous and unfailing cooperation of our Center colleagues. Dr. Timothy Francis of the Ships History Branch placed his draft narrative of *Constitution*'s operations at our disposal, and Robert Cressman and Mark Hayes, also from that office, provided useful comments. Volunteer Lillian Brodine and historians Gordon Bowen-Hassell and Dennis Conrad of the Early History Branch combined to transcribe and proofread some of the lengthier documents in the appendices. Edwin Finney, Jr., and Robert Hanshew of the Curator Branch's Photographic Section replied cheerfully to our many repeated requests for images from the Center's photographic collection. Davis Elliott of the Navy Department Library, Karin Haubold of the Navy Art Galley, Morgan Wilbur of Naval Aviation News, and intern Joshua Easterson of Gettysburg College, supplied timely and expert assistance in scanning many of the images that accompany this text, while Sandy Doyle, the Center's senior publications editor, gave helpful advice on preparing the final manuscript for publication.

The interpretations expressed herein are those of the authors alone, as are any errors of fact or interpretation.

Charles E. Brodine, Jr.
Michael J. Crawford
Christine F. Hughes

INTERPRETING OLD IRONSIDES

... as our Navy for a considerable time will be inferior in numbers, we are to consider what size ships will be most formidable, and be an overmatch for those of an enemy; such frigates as in the blowing weather would be an overmatch for double-deck ships, and in light winds evade coming into action.

Part I: Basic Level

Re-establishment of the Navy

Chronology

1783 Revolutionary War ends. Independence of the United States secured.

1785 Last ship of Continental Navy, frigate *Alliance,* sold.

Barbary corsairs begin seizing American merchantmen.

1789 U.S. Constitution gives Congress authority to establish a navy.

1793 Beginning of the wars of the French Revolution.

Barbary corsairs extend operations into the Atlantic Ocean.

1794 Congress authorizes six frigates.
Jay's Treaty between United States and Great Britain angers France.

1795 Peace treaty with Algiers.

1796 Congress authorizes completion of only three frigates.

1797 Launching of U.S. frigates *United States, Constellation,* and *Constitution.*

1798 XYZ Affair.
Navy increased.

Department of the Navy established.

Quasi-War with France begins.

U.S. Marine Corps established.

1799 U.S. frigates *Congress* and *Chesapeake* launched.

1800 U.S. frigate *President* launched.

1801 End of Quasi-War.

Narrative

After the United States won its independence, Congress, lacking the authority to impose taxes under the Articles of Confederation, was too weak to maintain more than a token armed force. The Continental Navy dissolved and the army dwindled to a mere seven hundred men.

The infant republic's military weakness convinced American nationalists of the necessity of adopting a new constitution that would increase the authority of the national government. The U.S. Constitution, ratified in 1789, gave Congress power to raise money to "provide and maintain a navy."

No new naval force was authorized, however, until the spring of 1794 when the clear necessity of defending the nation's seaborne commerce overcame congressional resistance. With the start of the wars of the French Revolution in 1793, warships of France and Britain began interfering with American trade with their enemies. Another threat to American commerce came from corsairs of North Africa's Barbary Coast. In 1785 Algerine corsairs made their first seizures of American vessels, holding passengers and crews captive, and in 1793 a truce between Portugal and Algiers opened the way for the latter's corsairs to cruise the Atlantic and imperil trade with much of Europe. Then the British prohibited all neutral trade with the French West Indies. In response to all these developments, on 27 March 1794 Congress authorized President George Washington to acquire six frigates, by purchase or otherwise.

Implementation of the 1794 naval legislation fell to the Department of War, headed by Secretary of War Henry Knox until 1795, Timothy Pickering from 1795 to 1796, and James McHenry from 1796 to 1798. Knox recommended the construction of new frigates designed to be superior to any vessel of that class in European navies.

The president approved six construction sites: Portsmouth, New Hampshire; Boston, Massachusetts;

Irish-born John Barry had a distinguished naval career during both the American Revolution and the Quasi-War with France. *Naval Historical Center photograph*

During the Quasi-War, citizens in American port cities subscribed moneys to fund the building of warships for the U.S. Navy. The 36-gun frigate *Philadelphia*, depicted here, was one such subscription ship. *Naval Historical Center photograph*

New York, New York; Philadelphia, Pennsylvania; Baltimore, Maryland; and Gosport (Norfolk), Virginia. At each site, a civilian naval constructor was hired to direct the work. Navy captains were appointed as superintendents, one for each of the six frigates. John Barry, last officer of the Continental Navy in active service, received commission number one as first officer in the new United States Navy.

The warships were still being framed when, in early 1796, word came of a negotiated peace with Algiers, at the cost of nearly one million dollars. The act authorizing the six frigates had called for a halt in construction in the event of peace with Algiers, but President Washington urged Congress to extend authorization to complete the six frigates. Congress approved the completion of only three of the frigates, and the frigate *United States* was launched at Philadelphia on 10 May 1797; *Constellation*, at Baltimore on 7 September 1797; and *Constitution*, at Boston on 21 October 1797.

In July 1797, the French government's disdain for American rights as neutral traders prompted Congress to authorize President John Adams to man and employ the three frigates. France had been America's major ally in the War of Independence, and without its assistance the United States may not have won independence. But the new government of Revolutionary France viewed Jay's Treaty, a 1794 commercial agreement between the United States and Great Britain, as a violation of the 1778 treaties between France and the United States. The French increased their seizures of American ships trading with their British enemies and refused to receive a new United States minister when he arrived in Paris in December 1796. In April of 1798 President Adams informed Congress of the infamous "XYZ Affair," in which French agents demanded a large bribe for the restoration of relations with the United States. Outraged, on 27 April 1798 Congress authorized the president to acquire, arm, and man no more than twelve vessels, of up to twenty-two guns each. Under the terms of this act several vessels were purchased and converted into ships of war.

The obvious need for an executive department responsible solely for, and staffed with persons competent in, naval affairs led Congress to pass a bill establishing the Department of the Navy. President Adams signed the act on 30 April 1798. Benjamin Stoddert, a Maryland merchant who had served as secretary to the Continental Board of War during the American Revolution, became the first secretary of the navy.

On 28 May 1798 Congress authorized the public vessels of the United States to capture armed French vessels cruising off the American coast, initiating the undeclared Quasi-War with France. On 11 July, the president signed the act that established the United States Marine Corps. On 16 July Congress appropriated funds to build and equip the three remaining frigates begun under the Act of 1794: *Congress,* launched at Portsmouth, on 15 August 1799; *Chesapeake*, at Gosport on 2 December 1799; and *President*, at New York, on 10 April 1800.

MJC

The Building of USS *Constitution*

Construction of the frigates authorized in 1794 fell to Secretary of War Henry Knox. Appointed to oversee both military and naval matters by President Washington, Knox had been consulting leading American shipbuilders and former officers of the Continental Navy on warship design and construction.

Strategic Vision Shapes Design

Among the secretary's consultants was Joshua Humphreys, a noted Philadelphia shipbuilder who had earned his reputation during the Revolution by skillfully converting merchantmen into warships for the Continental Navy as well as by constructing new men-of-war. In 1793 Humphreys, who had long pondered the problems of creating an effective American navy, summed up his thoughts in a letter to Knox:

> *... as our Navy for a considerable time will be inferior in numbers, we are to consider what size ships will be most formidable, and be an overmatch for those of an enemy; such frigates as in the blowing weather would be an overmatch for double-deck ships, and in light winds evade coming into action.*

By the spring of 1794, Knox and his associates reached a number of conclusions. The American frigates should be at least as big and powerful as any frigates then in existence. Hull construction should be as rugged as the technology of the day would permit; they should be as heavily armed as, or more heavily armed than, any single opponent they could not outsail; and finally, as long as even a modest wind prevailed, their rigging and hull form should give them speed to elude any enemy man-of-war or squadron that they did not have an excellent chance of defeating. Fulfilling these goals would not be an easy task but Knox had the advantage of starting with a clean slate, in that no pre-existing ships, manufacturing technique, or suppliers need inhibit the design and construction of the new ships.

On 28 June 1794 Knox officially appointed Humphreys to prepare plans for the frigates. Josiah Fox and William Doughty were hired soon after to assist him in translating his ideas into construction drawings. The three talented men freely exchanged ideas, and Humphreys continued to confer with such other experienced individuals as Captain John Barry and John Wharton. Though Humphreys is properly recognized for his key role in developing the 44-gun frigates, other minds deserve credit in the design.

Innovative Design Elements

To achieve their goals, the planners incorporated several innovative elements into the design of the new frigates.

The plans for the 44s specified a larger than customary vessel for a frigate of that rating, displacing 2,200 tons. One hundred and seventy-five feet between perpendiculars and forty-four feet two inches in the beam, with a depth of hold of fourteen feet three inches, they were twenty feet longer than their British contemporaries and thirteen feet longer than French 40-gun frigates. Being a little longer and slimmer than the standard frigates of the day contributed to *Constitution*'s speed and maneuverability.

Extensive use of live oak succeeded in making the hull strong and durable. The hull consisted of three layers: outer and inner horizontal layers (planking) of white oak, and a center, vertical layer (frames) of live oak. Live oak

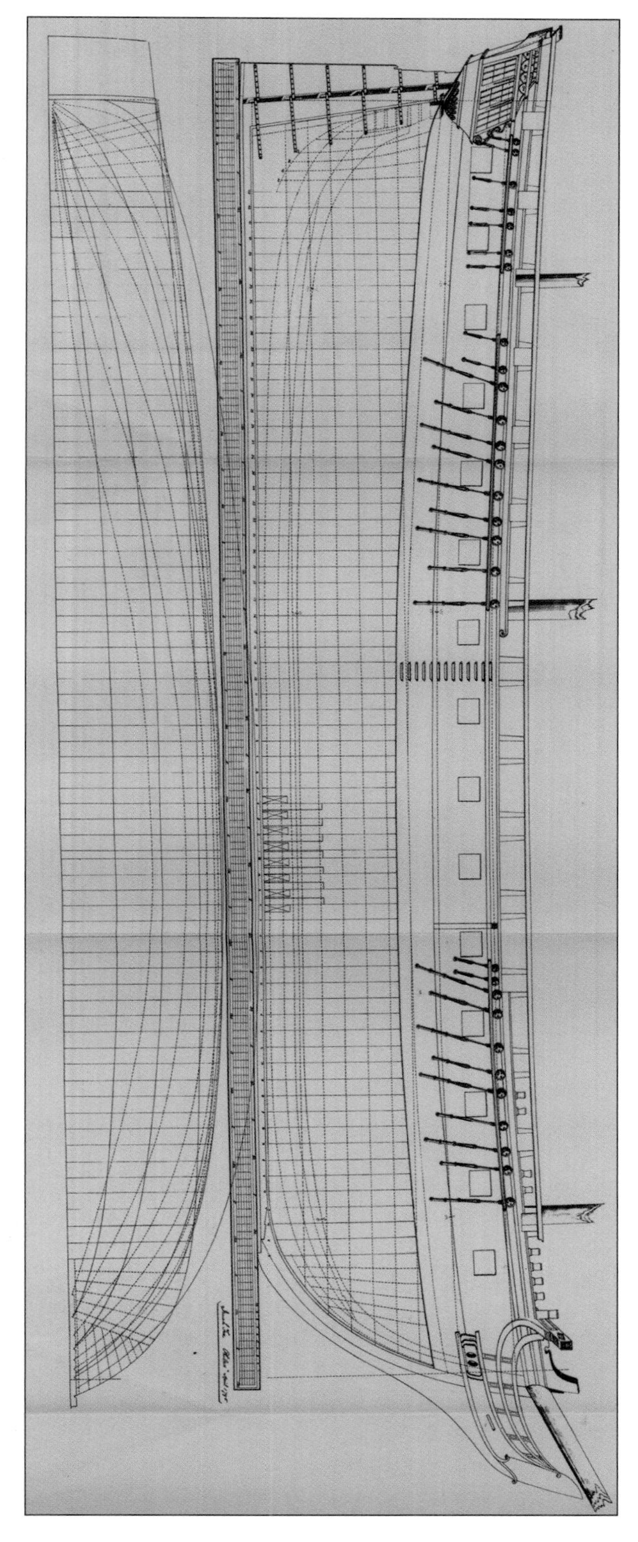

Naval Constructor Josiah Fox's sheer and half breadth plan for a 44-gun frigate transformed Joshua Humphreys's concept for USS *Constitution* into a draft design. *Register of Officer Personnel: United States Navy and Marine Corps and Ships' Data, 1801–1807, plate IV*

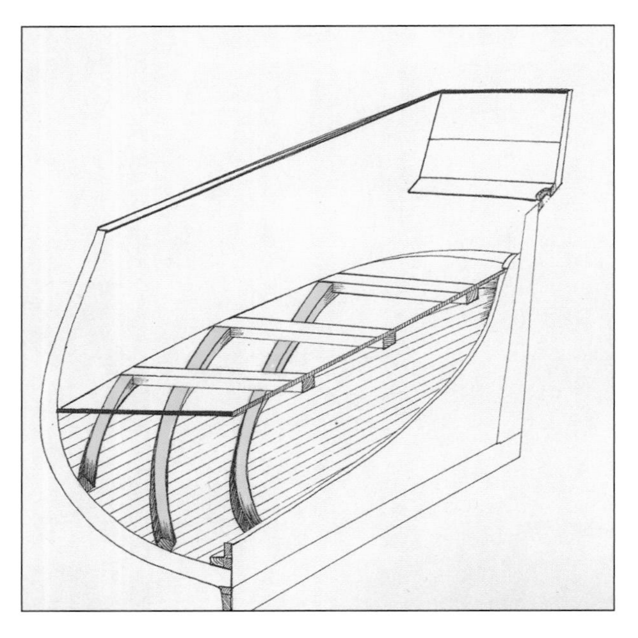

Diagonal riders (in yellow) added longitudinal strength to *Constitution*'s hull. *Naval Historical Center photograph*

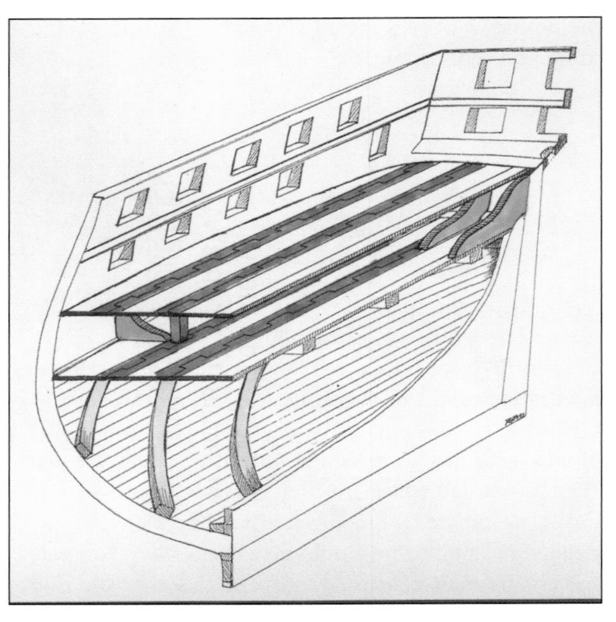

Lock-scarfed or notched planking (in red) reinforced the ship's decks. *Naval Historical Center photograph*

is about five times denser than other oak woods, and, at the close of the eighteenth century was only available in the southeastern United States. The live oak for *Constitution* came from the swampy coastal forests of Georgia. Additionally, the live oak frames were placed only two inches apart, instead of being spaced from four to eight inches, as were the frames of British and French frigates.

Copper pins made in Boston by a local coppersmith, Paul Revere (famous for his midnight ride), and about 150,000 wooden pegs called "treenails" (pronounced "trunnels") held the hull together. This unusually strong hull had an average thickness of twenty-one inches, and was twenty-five inches thick at the waterline. The hull's underside was plated with copper sheets imported from Great Britain. Tarred paper known as "Irish felt" was placed between the hull and the copper sheeting.

To enable the large frigates to carry a main armament of heavy 24-pounders without "hogging" (that is, sagging at the bow and stern) Humphreys introduced a system of diagonal riders. Within each side of the hull, six thick curved timbers ran from the keel to the gun deck, transferring the downward pressure of the cannons' weight to the center, rather than to the ends, of the vessel.

Thick beams, shaped roughly like inverted letter Ls and called "standard knees," spaced along the length of the berth deck, transferred the weight of the long guns on the gun deck to the diagonal riders.

Mutually reinforcing design features made possible the placement of cannon along the entire length of the nearly continuous upper deck, called the spar deck. Tying together the spar deck's planking, which ran the length fore and aft, with lock-scarfs, and fastening them to the timbers at the fore and aft stiffened the ship and further countered the tendency that large vessels have of hogging. Lock scarfing refers to the way that the deck planks are cut. Several deck planks on each deck are notched along their lengths so that they interlock, like puzzle pieces, with the adjoining planks. Heavy stanchions on the deck below supported each gun.

In summary, the final design for the 44s incorporated several innovative elements that gave the new frigates increased strength: additional length; extensive use of live oak; diagonal riders; lock scarfing; and standard knees.

Constitution Built and Launched

To speed construction, and to distribute the economic benefits of the project throughout the states (a policy well known even then), President Washington directed the newly authorized warships be built in each of six American shipbuilding centers. He assigned one of the 44-gun frigates to Boston, with the other five warships built in Baltimore, New York, Norfolk, Philadelphia, and Portsmouth.

On 2 July 1794 Henry Jackson was engaged as the Boston naval agent, responsible for administering the local purchasing process and for hiring skilled artisans and laborers. Colonel George Claghorne was appointed constructor, to direct the shipyard work force. Finally, Knox named Captain Samuel Nicholson, a veteran of the Continental Navy, as the prospective commanding officer of the frigate.

As wooden warships of the time were ornamented with hand-carved woodwork at stem and stern, Timothy Pickering, who took Henry Knox's place as secretary of war at the beginning of 1795, felt that the decoration of each ship should be related to its name. Since such work took time, Pickering sent a list of ten suggested names to President Washington on 14 March 1795. One of the names the president selected was *Constitution*, and this name was allotted to the Boston frigate.

Constitution was laid down at Edmund Hartt's Boston shipyard during the summer of 1795. Edmund, one of the three Hartt brothers who owned the shipyard, became the yard foreman and worked in cooperation with Colonel Claghorne.

In early March 1796, word reached the shipyard in Boston that the dey of Algiers had signed a peace treaty with the United States. The peace with Algiers, under section 9 of the Naval Act of 1794, should have ended *Constitution*'s life before it began. Before suspending work on the frigates, however, President Washington suggested that abrupt cancellation of the warships would be wasteful and disruptive of economic life, and asked Congress to reconsider the matter. It was also well understood in Congress that the completed frigates were destined for use in the Mediterranean, as the Barbary States could not be trusted in the long run to keep the peace. On 20 April 1796 the nation's lawmakers authorized the president "to continue the construction and equipment (with all convenient expedition) of two frigates of forty-four [*United States* and *Constitution*], and one of thirty-six guns [*Constellation*], any thing in the act, entitled 'An act to provide a naval armament,' to the contrary notwithstanding."

Work on the frigates continued that year, with the Portsmouth yard ordered to send any live oak in their possession to help finish *Constitution*. Although building costs proved to be greater than originally estimated, Congressional advocates of the Navy managed to persuade their more reluctant colleagues to approve additional appropriations needed to finish the frigate. They were aided by the spread of war in Europe and French victories in Italy. On 1 July 1797, another act authorized the president, "should he deem it expedient, to cause the frigates *United States*, *Constitution*, and *Constellation* to be manned and employed."

Shortly before noon on 20 September 1797, Claghorne ordered the launching of *Constitution*. Unfortunately, the warship slid only twenty-seven feet toward the water before coming to rest, the launch way having settled in the mud just enough to bring *Constitution* to a standstill. The yard gang drove wedges to raise the ramp, and tried again on the 22nd; but *Constitution* moved only thirty-one more feet, still short of the water. The failures delighted opponents of the Navy, with Jeffersonian newspapers and pamphlets gleefully trumpeting the faux pas, hoping the frigate—which they saw as a symbol of overweening Federal power—would never get to sea. On 18 October the Republican newspaper editor and poet Philip Freneau celebrated the event with an ode, "To the Frigate *Constitution,*" urging the ship to remain high and dry.

Madam! – Stay where you are, 'Tis better, sure, by far
Than venturing on an element of danger
Where heavy seas and stormy gales
May wreck your hulk and rend your sails
Or Europe's Black-guards treat you like a stranger,

When first you stuck upon your ways
(Where half New England came to gaze)
We antifederals thought it something odd
That where all art had been display'd,
And even the builder deem'd a little god,
He had not your ways better laid.

O frigate Constitution! stay on shore:
Why would you meet old Ocean's roar?

Freneau's poem continued on in like vein for several more stanzas, clearly demonstrating the significant anti-Federalist hostility that existed against rebuilding American military capabilities, particularly when it came to the more expensive warships. Better times though, awaited the frigate, not to mention better poetry.

On 21 October 1797 the tide again approached its maximum height and one of *Constitution*'s guns—still ashore to save on weight—fired to announce the frigate's third attempt to launch. Claghorne had increased the slope of the ways, and this time the ship finally entered the water in proper fashion. While Captain Nicholson stood on *Constitution*'s deck as the ship floated in its element, Captain James Sever stood at the heel of the frigate's bowsprit, broke a bottle of choice Madeira across its bow, and ceremonially bestowed the name *Constitution* on the new ship.

MJC

Constitution's Guns

The two most important types of cannon in the American fleet during the War of 1812 were long guns and carronades. Both of these categories of gun were classified according to the weight of shot they fired. That is, a 9-pounder cannon fired a nine-pound ball, a 42-pounder carronade fired a forty-two pound ball, and so on. American long guns ranged in size from 9- to 24-pounders while carronades ranged in size from 18- to 42-pounders. In the blue water Navy of which

Constitution was part, schooners (such as *Carolina*), brigs (such as *Argus*), and sloops of war (such as *Wasp*) mounted carronades as their primary armament while all U.S. frigates (except *Essex*) carried mixed batteries of long guns and carronades. Typically, American warships, and those of the Royal Navy too, mounted larger numbers of guns than they were officially rated to carry. USS *Chesapeake,* for example, though designated a 36-gun frigate, bore forty-nine guns at the time of its capture in 1813. *Constitution,* a 44-gun frigate, carried between fifty-two and fifty-five cannon during the War of 1812.

Long guns had served as the chief armament in sailing warships since the seventeenth century. This type of cannon, so named for its long barrel, was mounted on wheeled, wooden carriages and secured to a ship's deck with ringbolts, blocks, and tackle. Originally *Constitution* carried only guns of this class, 12-, 18-, and 24-pounders. During the War of 1812 the frigate carried thirty 24-pounders on its main or gun deck. These guns, cast by Samuel Hughes of Cecil Furnace, Maryland, were made of iron, weighed (exclusive of carriage) 5,824 pounds each, and had a range of 1,200 yards. Long-barreled cannon were more accurate and had a greater range than other kinds of naval ordnance, enabling a ship to engage its foe at a distance. Almost all warships, including *Constitution,* carried one or more long guns on their weather deck, mounted at the stern or bow, to serve as chase guns. In this way, gunfire could be delivered against enemy vessels whether the ship was in flight or pursuit. Isaac Hull employed long guns in this fashion against a British squadron during the "Great Chase," 16-19 July 1812.

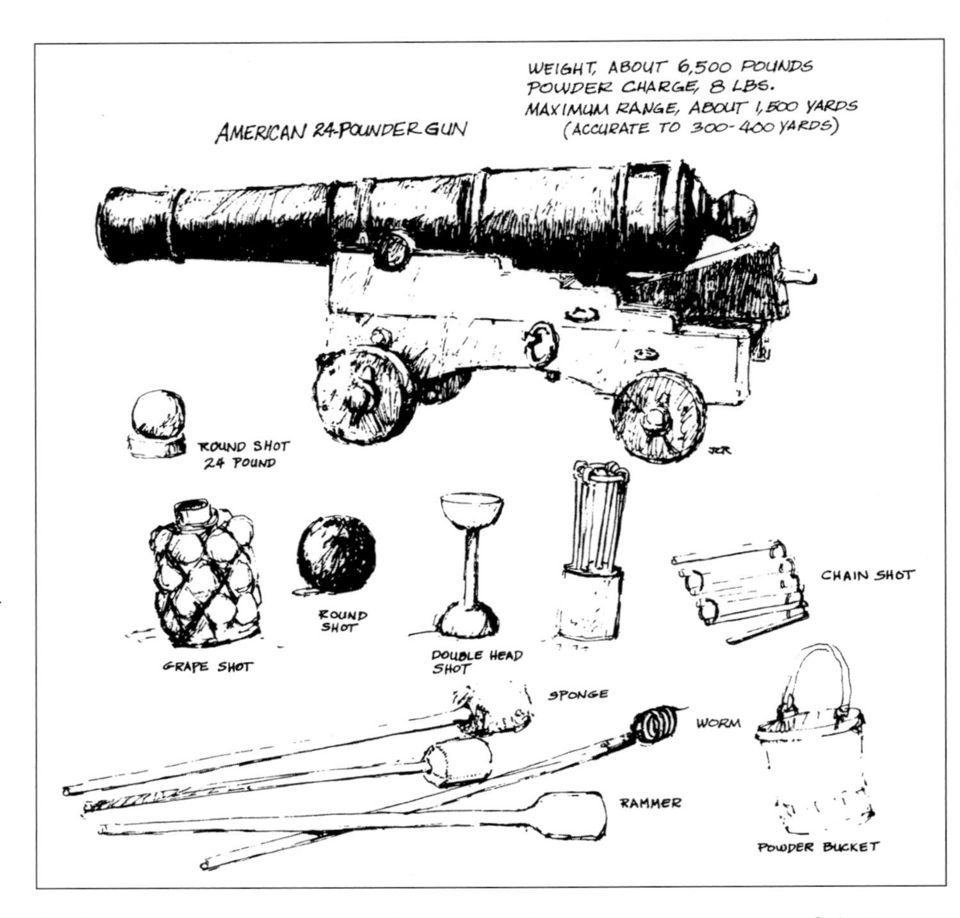

Constitution's main battery consisted of thirty 24-pounder long guns. These guns were of a heavier caliber than those typically carried in British frigates in 1812, giving *Constitution* a decided edge over its Royal Navy counterparts. *Old Ironsides, U.S. Frigate Constitution: An Essay in Sketches*

The carronade, an innovative gun of British design, did not enter the arsenal of sailing navies until the last decades of the eighteenth century. Josiah Foxall's Columbia Iron Works of Pennsylvania cast *Constitution*'s carronades. Each gun was made of iron, weighed 1,918 pounds (exclusive of carriage), and could deliver shot with effect at up to 400 yards. During the war, *Constitution* mounted between twenty and twenty-four 32-pounder carronades on its spar deck. Carronades differed from long guns in a number of significant ways. They had lighter, shorter, more thinly molded barrels than those of long guns. They were also mounted on carriages in which a slide rather than blocks and tackles absorbed the gun's recoil. But the most impressive attribute of carronades was their hitting power. Round shot fired from these guns wreaked havoc on opposing vessels, knocking large gaping holes in their hulls and bulwarks and generating a shower of secondary, splinter projectiles that maimed and killed personnel. Just how formidable these "iron Attila's," as author Herman Melville dubbed them, could prove in a ship to ship action is evidenced in the 1813 engagement between the U.S. sloop of war *Hornet* and H.M. brig-sloop *Peacock*. In a combat lasting less than a quarter of an hour, *Hornet*'s 32-pounder carronades reduced *Peacock* to a sinking state, killing and wounding more than a quarter of its crew.

Despite its ability to deliver destructive firepower, the carronade had one serious shortcoming in action—a range that was roughly one third that of long guns of the same caliber. A ship armed solely with carronades was thus limited tactically, able to engage its foe in gunnery at close quarters only. Likewise, a vessel mounting long guns held a distinct advantage in battle over a ship armed with carronades, as it could then engage its foe from a distance outside the range of the latter's guns. The most spectacular example of how long guns provided the winning edge over carronades in combat was the victory of HMS *Phoebe* and *Cherub* over the U.S. frigate *Essex* in 1814. *Constitution*'s defeat of *Cyane* and *Levant* in February 1815 also illustrated the limitations of carronades in battle. In that action, Captain Charles Stewart took advantage of adroit seamanship and *Constitution*'s superiority in long guns to hammer away at both British warships while remaining out of reach of their carronades. As a result, the British were never able to bring their heavier combined broadside to bear on *Constitution,* and the American frigate chalked up another impressive victory against the Royal Navy.

The gunner was the officer charged with the care and maintenance of *Constitution*'s guns. He held his warrant by Navy Department appointment or through an acting appointment issued by the captain. A number of petty officers, including the gunner's mate, quarter gunners, and ship's armorer, assisted the gunner in the management of his department. Keeping the frigate's carronades and long guns in operating condition was one of the gunner's most important and challenging duties. This meant repairing ordnance damaged in battle or during gunnery exercise; cleaning guns fouled by repeated firings; and keeping cannon free of rust. Inspecting the ship's powder stores was another vital task entrusted to the gunner. Failure to keep the gunpowder free from moisture or well turned could ruin its explosive properties, thereby contributing to the ship's defeat in battle. The gunner was also responsible for the tools gun crews used to move, aim, load, and clean the guns during firing (handspikes, quoins, rammers, swabs, and worms). He was required to maintain accurate, up-to-date inventories of all the stores in his department including gunpowder, cartridges, priming tubes, cannon shot, etc. In battle, the gunner was stationed in the frigate's magazine to supervise the making and distribution of cartridges for the guns. It is likely that *Constitution*'s gunner assisted in the design and construction of a furnace the ship carried for heating cannon shot. This furnace, the inspiration of Captain Stewart, was capable (according to British intelligence) of heating forty-five shot in fifteen minutes.

The guns in *Constitution* fired several kinds of projectiles: solid or round shot for damaging hull, masts, and spars; chain, bar, and star shot for destroying sails and rigging; and canister and grape shot for killing personnel. Fire was delivered in one of two ways: direct, with guns elevated to deliver shot on target without grazing, and, ricochet, with guns leveled to skip shot off the surface of the water at a target. Ricochet fire, though less destructive, could strike targets at twice the range of direct fire. Battery fire was given in one of three ways: independently, in succession, or concentrated (broadside). Broadside was the least preferred way of delivering fire because the simultaneous discharge of one side of a ship's guns violently shocked a vessel's structure. The powder charges for a ship's guns were usually wrapped in flannel, though paper and lead wrappings were sometimes used. A full service charge weighed roughly one third of the projectile being fired. Thus an 8-pound charge was used for a 24-pound shot. Smaller charges were used as the range between combatants decreased. *Constitution*'s 24-pounder long guns used 8-, 7-, and 6-pound service charges, while 4-pound charges were used in live fire drill. One interesting fact about the guns aboard American ships is that the crew sometimes named them. In the frigate *Chesapeake,* guns bore such names as: Raging Eagle, Liberty or Death, United Tars, and Jumping Billy.

Three classes of British frigates patrolled American waters between 1812 and 1815, those of 32-, 36-,

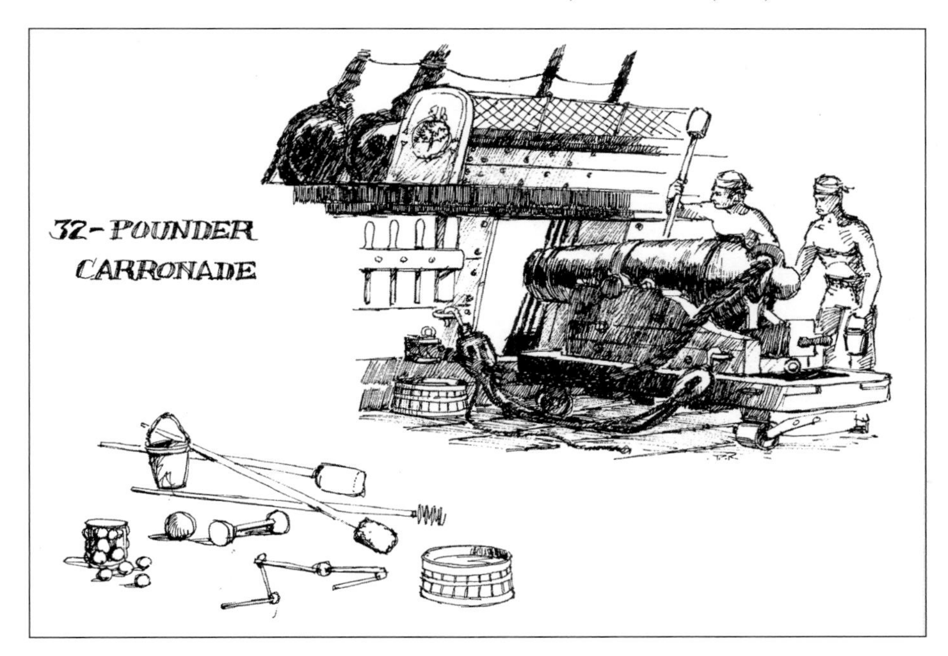

Constitution carried up to two dozen 32-pounder carronade guns on its spar deck during the War of 1812. Also known as "smashers," these guns provided devastating firepower in actions fought at close range. *Old Ironsides, U.S. Frigate Constitution: An Essay in Sketches*

and 38-guns. The two frigates *Constitution* engaged in 1812, *Guerriere* and *Java,* were of the 38-gun class. Like Old Ironsides they carried a mixed battery of cannon—long guns and carronades. And like *Constitution,* they mounted 32-pounder carronades on their upper deck. There the similarities ended, for the British vessels carried 18-pounder long guns on their gun decks while the American frigate boasted 24-pounders. Because of this disparity in force, *Constitution* enjoyed an advantage of 3 to 2 in weight of metal thrown in broadside. The American frigate's superiority in armament enabled it to reduce both *Guerriere* and *Java* to dismasted, sinking wrecks. Because of these humiliating defeats, the British Admiralty sought to nullify the American advantage in heavy armament by developing a larger gun to replace the 18-pounders then in their frigates. This initiative resulted in a new 24-pounder gun designed by Sir William Congreve. But the war ended before the Royal Navy had an opportunity to test its new gun against the American navy. Thus *Constitution* and its sister frigates still reigned supreme as the most powerful ships of that class afloat.

CEB

An outboard view of *Constitution*'s starboard gun deck battery.
U.S. Navy photograph

A Ranked Society: Command Structure in Old Ironsides

Rank was the great organizing principle in Navy warships during the age of sail. It defined what one did, how one dressed, and where one messed and slept on board ship. It influenced how one was disciplined and determined what one was paid. It established one's standing in the ship's company and governed social and professional relations between hands. Indeed, not one aspect of shipboard life was unaffected by notions and customs of rank. The following text will briefly sketch the various ranks and associated duties held by the men of *Constitution* during the War of 1812. Attention is given here to the officers and men entrusted with the professional management of the ship. The duties of several of the frigate's civil officers (purser, surgeon, and surgeon's mate) and of its complement of Marines are described in lesson plans elsewhere and in the Navy Department Regulations in the appendices.

The captain stood at the head of *Constitution*'s crew. The safety of the ship, in port and at sea, was his paramount responsibility, and his authority over the frigate, and those serving in it, was unqualified. "The captain's word is law," declared novelist Herman Melville. "He never speaks but in the imperative mood. When he stands on his Quarter-deck at sea, he absolutely commands as far as eye can reach. Only the moon and stars are beyond his jurisdiction." The captain delegated responsibility for the day-to-day running of the frigate to his executive officer, the first lieutenant, exercising only a general superintendence over the ship's operations. In battle, hazardous weather, or like situations, however, he personally directed the ship. Overseeing the professional training of the ship's junior officers, especially the midshipmen, was among the captain's more important duties. He did this through a combination of mentorship, instruction, and personal example. Because Navy captains cultivated an aura of authority, reserve, and even severity as part of their command persona, they were, in many ways, unapproachable figures aboard ship. *Essex*'s commander, David Porter, captured this sense of isolation when he described a ship's captain as "a solitary being in the midst of an ocean."

The next in line of authority on board *Constitution* was the first lieutenant, the senior officer of that grade among the ship's company. As second-in-command, the first lieutenant had extensive and demanding duties. He was responsible for inspecting the frigate daily, noting any deficiencies in the state of the vessel or crew. He was the person through whom all communications

Navy officers and seamen in full dress uniform, 1812–1815, based on contemporary drawings and descriptions. Clothing was the most visible indicator of one's rank within the naval hierarchy. *Company of Military Historians®*

to and from the captain passed. He had charge of the deck when all hands were summoned for particular operations or evolutions. In battle he commanded the quarterdeck gun division and the ship's boarding parties. He also drew up the ship's station, watch, and quarter bills. Because his duties consumed so much time, the first lieutenant (like the captain) stood no watch.

The ship's remaining lieutenants, usually five in number and ranked according to seniority, assisted the executive officer in running the ship. One of their chief duties was to stand watch as officer of the deck during which time they were responsible for the safety and proper management of the ship. This included keeping the men at their duty, receiving reports from the various departments, making entries in the vessel's logbook, sounding the well, and mustering the night watch. Each lieutenant was also given direction of one of the frigate's gun divisions and supervised the combat training of the Sailors attached to it.

One step further down Old Ironsides' command ladder stood its sailing master and midshipmen. These officers held their warrants by appointment of the Navy Department, or by acting appointment of the frigate's commander. The sailing master was the most senior of these two grades and assisted in the navigation of the ship. Because he was required to keep the ship in proper sailing trim, the master had charge of the ship's ballast and stowage of all of its stores. He held the keys to the spirit room, inspected the galley for cleanliness, and bore overall responsibility for appropriate entries (noon reckonings, wind, speed, temperature, course changes, ship sightings, etc.) in the ship's log. Navy Regulations assigned no particular duties to midshipmen as they were considered a class of apprentice officers whose time was best spent mastering the fundamentals of seamanship and command, in preparation for assuming, one day, commissioned rank. To this end, they were required to keep journals, receive instruction in navigation, study naval tactics, and learn every skill requisite to service in a man-of-war. Midshipmen kept watches, helped exercise the men at the guns, commanded work details, and were given charge of prize vessels.

The four remaining warrant officers in *Constitution* (the boatswain, gunner, carpenter, and sailmaker) occupied an inferior status in the ship's company relative to the sailing master and midshipmen. This is because the duties they performed on board Old Ironsides were regarded as manual labor, something inconsistent with the calling of a gentleman officer. The departments of these four men carried out vital upkeep and repairs that maintained the frigate's seaworthiness. The boatswain was responsible for the ship's rigging. His usual station in *Constitution* was the forecastle, where he supervised the working of the ship. He helped rig the vessel for action and repair battle damage. He was also the chief of the crew, and as such, responsible for turning out the watch and summoning all hands to perform various evolutions. The boatswain carried a rattan cane, which he used to enforce discipline and attention to duty among the crew. The carpenter was responsible for the wooden structure of the ship—hull, masts, and yards. His duties also embraced the care of *Constitution*'s boats. The sailmaker cared for over an acre of canvas comprising the ship's forty-odd sails, while the gunner maintained the frigate's ordnance, munitions, and small arms. All four officers were obliged to report regularly in writing and orally to the first lieutenant on the state of their departments and to keep records of the stores entrusted to their care.

Next below on the frigate's chain of command were *Constitution*'s petty officers. The majority of these were mates or assistants to the following warrant officers: the sailing master (master's mates and quartermasters); the boatswain (boatswain's mates and boatswain's yeomen); the gunner (gunner's mates, quarter gunners, and armorer); the carpenter (carpenter's mates, carpenter's yeomen, and cooper); and the sailmaker (sailmaker's mates). These men performed duties which, by and large, appertained to the department in which they served. For example, the master's mates and quartermasters assisted the sailing master with navigating and conning of the ship; the gunner's mate and quarter gunners assisted the gunner in keeping the guns and carriages in their proper order; the carpenter's mates assisted the carpenter in examining and repairing the ship's spars; and so on. Four additional petty officers attached to the frigate were the purser's steward, the master-at-arms, the cook, and the coxswain.

At the bottom of *Constitution*'s hierarchy resided the common Sailor, who comprised some 70 percent of the ship's company. These personnel were rated on the frigate's books according to their nautical skill and capacity. Boy was the lowest rating recognized in naval law in 1812. As its name implies, this class of sailor was composed of young men who either lacked knowledge of seamanship, or, who were too small and weak to manage more sailor-like tasks aboard ship. Adult men who were novices to the mariner's world were rated as landsmen. Melville described such greenhorns as "inveterate 'sons of farmers' with the hay-seed yet in their hair." Although the names of personnel entered as landsmen appear in Navy muster rolls as early as 1798, it was not officially recognized as a rating in Navy pay tables until 1838. Men who had had enough sea time to master the rudiments of helmsmanship and handling sails aloft were rated as ordinary seamen. The most capable Sailors aboard ship were rated as seamen (sometimes designated in muster rolls as able seamen). These were men of long experience at sea, well versed in navigation, handling sails, helmsmanship, and rigging work. Seamen, one naval chaplain observed in 1832, have a great "pride of profession, entertaining the utmost contempt for all who do not know what salt water and heavy gales are." Charles Stewart carried an exceptionally high number of seamen on *Constitution*'s rolls during his first and second wartime cruises, up to 85 per cent more than the number authorized for 44-gun frigates. He may have done this to compensate for the loss through transfer of many veteran crewmen in the spring of 1813, entering large numbers of skilled Sailors as replacements.

All of *Constitution*'s Sailors, boys to seamen, participated in regular combat-related drills: guns, small arms, boarding, etc. They also spent considerable energies keeping the ship's spaces clean and well arranged, although this latter duty was not shared equally among the ratings. In his novel *White-Jacket,* depicting life in an American man-of-war in the late 1840s, Herman Melville wrote that the dirtiest and most detestable jobs aboard ship such as "attending to the drainage and sewerage below hatches" and policing the ship's "chicken-coops, pig-pens" fell to landsmen. Some of the typical chores assigned to boys in *Constitution* included serving as officer's servants, messenger boys, galley cooks, and powder monkeys. During evolutions requiring the handling of sails, crewmen were assigned to one of six divisions aboard ship. Landsmen were appointed to serve in the waist (waisters) and on the quarterdeck (afterguard) to handle the work of hauling on cables, ropes, and lines, a duty that required more brute strength than sailorly skill. The most active and dexterous ordinary seamen and seamen (and sometimes boys) manned the three different tops (fore, main, and mizzen). A cadre of the ship's most experienced seamen, known as sheet anchormen, served in the forecastle. Their job was to handle the foreyard, anchors, and all sails on the bowsprit. In each of these divisions (forecastle, waist, afterguard, foretop, maintop, mizzentop) the smartest, most alert men were chosen to be division captains.

CEB

The Men of *Constitution*

Several attributes contributed to *Constitution*'s formidable character as a combat vessel: a hull design that yielded superior speed under sail; an armament that delivered punishing firepower at close and long range; and a heavy, plank-on-frame construction that provided greater protection against enemy shot and shell. Yet another component of *Constitution*'s strength was its large complement of men. More than 440 men sailed on each of the ship's wartime cruises, better than a third more personnel than carried in the 32, 36, and 38-gun British frigates stationed off the American coast. This larger number of men gave *Constitution* a significant edge over opposing vessels in wartime operations. In battle, it meant additional hands to serve the ship's guns, manage its sails, assist in damage control, participate in boarding actions, and replace men fallen in combat. It gave the ship's commanders greater flexibility in manning and sending in prizes. It also provided the ship with a larger reserve for replacing crewmen lost through sickness, accident, or desertion.

In 1797, Congress authorized 44-gun frigates to carry the following complement: 9 commissioned sea officers, 14 warrant officers, 28 petty officers, 253 seamen and ordinary seamen, and 60 Marines, for a total of 364 men. In practice, *Constitution* and its sister 44s carried far larger crews than this, though with official sanction. What Navy administrators considered an appropriate strength for heavy frigates was never set forth in Departmental orders or instructions. However, some notion of the Department's thinking on this subject is reflected in an 1807 estimate for *Constitution* that details a crew of 420 men, while an 1816 estimate for a 44-gun ship calls for a crew of 450. Although commanders were usually given great latitude in how they manned their ships, departure from accepted norms could result in censure. In the fall of 1813, Charles Stewart drew Commodore Bainbridge's criticism for shipping excessive numbers of petty officers and seamen (the most highly skilled class of sailor). Secretary Jones also reprimanded the frigate commander for carrying an extra lieutenant on the ship's rolls. (See p. 133, Authorized Strength, 44-Gun Frigate.)

The ships of the American fleet obtained their crews in one of two ways, drafts and naval rendezvous. Drafts, or transfers of men from one ship to another, were the quickest way to fill out a vessel's complement. The Navy Department employed this practice extensively to supply U.S. warships on the Great Lakes with Sailors. The more traditional method of shipping a crew was to recruit Sailors at naval rendezvous (recruiting centers) opened in seaport towns and cities. This is how *Constitution* acquired most of its enlisted men. Officers, on the other hand, received their ship assignments from the Navy Department. As a general rule, captains did not enjoy the privilege of selecting their own officers. They inherited the wardroom and steerage officers of the ships of which they had been given command. A captain could solicit the appointment of particular officers to his vessel and the Navy secretary often, though not always, honored these requests. When Secretary Jones disallowed the transfer of one of Charles Stewart's former lieutenants in *Constellation,* Jesse Wilkinson, to Old Ironsides, Stewart vented his frustration at not having "at least one officer with whom I am acquainted and of whose talents I have some knowledge." Jones placated Stewart by ordering *Constellation*'s William B. Shubrick to *Constitution* in place of Wilkinson.

Despite the lack of detailed naval personnel records for the War of 1812, it is possible to offer some general conclusions about the enlisted men who served in *Constitution* at that time. First, the majority of the ship's crewmen were young (between the ages of sixteen and twenty-nine), single, and native-born Americans. Second, the ship's company included a number of British-born Sailors, veterans of Royal Navy service. Isaac Hull discharged a number of such men in July 1812, as they feared being hanged for desertion should *Constitution* fall prize to enemy warships. Third, black Sailors formed an important part of the frigate's complement, as they did throughout the rest of the fleet. It is estimated that black seamen provided 15-20 percent of the Navy's manpower at this time. Isaac Hull reckoned blacks *Constitution*'s best fighters in the engagement with *Guerriere,* observing they "fought like devils," were completely fearless, and seemed "possessed with a determination to outfight the white Sailors." Finally, it is noteworthy that numbers of *Constitution*'s Sailors came to the ship with important skills learned in trades ashore, such as carpentry and blacksmithing. These men, observed *Constitution*'s first lieutenant Charles Morris, proved invaluable in repairing damages to the frigate while under way.

While much is known about *Constitution*'s three wartime commanders, Isaac Hull, William Bainbridge, and Charles Stewart, little is known about the majority of officers who served under them. Reviewing the service records of the frigate's lieutenants and midshipmen, however, provides some intriguing insights into the changing nature of leadership in Old Ironsides during the war. It is notable that *Constitution* began the war led by a reasonably experienced cadre of officers. Four of the ship's six lieutenants had served under Hull for two years, while eight of its fifteen midshipmen had made at least one cruise under the Connecticut captain's

I, Alexander Eskridge of Leesburg Virginia appointed a midshipman in the navy of the United States do solemnly swear to bear true allegiance to the United States of America, and to serve them honestly and faithfully against all their enemies or opposers whomsoever; and to observe and obey the orders of the President of the United States of America, and the orders of the officers appointed over me, and in all things to conform myself to the rules and regulations which now are or hereafter may be directed, and to the articles of war which may be enacted by Congress, for the better government of the navy of the United States, and that I will support the constitution of the United States. City of Norfolk

SWORN BEFORE ME, this 22d Feby 1812
[illegible] Mayor

Alexander Eskridge

This oath of allegiance, sworn to by Midshipman Alexander Eskridge of *Constitution*, was required of all individuals entering the Navy. Eskridge served the entire war in Old Ironsides, a distinction shared by only a handful of the frigate's 1812 company. *Record Group 45, Letters Received Accepting Appointments as Midshipmen, 1809–39, National Archives and Records Administration* command. Though few of these men had witnessed combat, their leadership and ability enabled Hull to lead what had largely been a green crew to victory on 19 August 1812 over *Guerriere*. With only a few changes in personnel, this same collection of officers helped William Bainbridge conquer *Java* four months later. Not surprisingly, *Constitution*'s combat successes resulted in a flurry of promotions. Three lieutenants, Charles Morris, George Parker, and George C. Read, were awarded independent commands (*Adams, Siren,* and *Vixen* respectively), while eight midshipmen were elevated to lieutenancies in *Adams, Siren, Frolic,* and *Wasp*. Overall, this represented an important and striking contribution to the command structure of the fleet.

But what was good for the fleet was not good for *Constitution* and it was Charles Stewart's misfortune to assume command of the frigate (July 1813) at a time of significant turnover in officers and men. Only one lieutenant, Beekman V. Hoffman, and five midshipmen remained from Hull's command. Two of the latter, Shubael Pratt and James Delany, would prove so troublesome that Stewart sought to have them put out of the ship. On the eve of *Constitution*'s final cruise of the war, the overall impression of its officers is one of youth and inexperience compared to the glory days of 1812. Nine of the ship's seventeen midshipmen had received their warrants since the commencement of the war (18 June 1812) and two of its newly commissioned lieutenants had never stood watch at sea. It is a tribute to Charles Stewart's command abilities that he led such a company of men to victory over *Cyane* and *Levant*.

A number of factors combined to erode *Constitution*'s crew size, so keeping the ship fully manned was a constant challenge to its commanders. Deaths, due to combat, accidents, and sickness, accounted for small, but significant losses of crewmen. Of the fifty-two wartime deaths noted in official records, twenty-seven were combat related. Many more men were temporarily lost to incapacitating illness or wounds sustained in battle. At times, the names of as many as forty-five men appeared on the ship's sick list. By far the most vexing loss of men was owing to desertion. The ship's surviving muster rolls, covering the period of Hull's and Stewart's captaincies, record 108 desertions. Seventy percent of these occurred during the first six months of Stewart's command. Indeed, during the summer months of 1813, the ship seemed to hemorrhage men, with forty-six Sailors stealing away to freedom. Additional reductions in the frigate's rolls came through drafts made on its crew. In April 1813, Commodore Bainbridge transferred one hundred fifty of *Constitution*'s Sailors to the squadron on Lake Ontario. Follow-up drafts were sent to the Lakes again in June. A final drain on *Constitution*'s manpower occurred through the regular discharge of men whose two-year terms of enlistment were up.

Shortages of men in *Constitution* had a profound effect on the ship's operational readiness. For example, they delayed Hull's departure from the Chesapeake in 1812 and Stewart's sailing from Boston in 1813 and 1814. Combined with the significant turnover in personnel that took place in 1813, they contributed to a decline in the crew's efficiency, competency, and discipline. Fortunately, Old Ironsides was blessed with three experienced and capable commanders, with the knowledge and ability to mold men into a battle-ready force.

On presenting his sword to Isaac Hull following the capture of *Guerriere,* Captain James Dacres praised the fighting prowess of the American tars in battle. *Constitution*'s men, he declared, fought "more like tigers than men. I never saw men fight so. They fairly drove us from our quarters." In paying homage to Old Ironsides' men in this way, Dacres acknowledged a hard-learned lesson in which his brother officers would be fully schooled by war's end. That is, that the strength of the United States Navy lay not only in the size, speed, and power of its ships, but also in the exceptional fighting spirit and patriotism of its men.

CEB

Constitution's Marines

On 30 June 1798, Congress authorized the creation of the Marine Corps. As its founding legislation made clear, the Marines were meant to serve either at sea in the new Navy's ships, or ashore in coastal forts and garrisons. The administration, training, and discipline of the Corps reflected this duality in mission—as both a land and sea service. For example, the secretary of the navy exercised responsibility for the overall direction of the Corps, while a major (later lieutenant colonel) commandant, whom a small cadre of staff officers bearing military titles (adjutant, paymaster, quartermaster, etc.) assisted, executed the day-to-day running of Marine Corps business. When serving afloat Marines came under Navy discipline; ashore the Army's Articles of War prevailed. And while Marines might train to fight in ships at sea, the guns they used and the drill manuals they trained with were the Army's. Writing two decades after the War of 1812, Navy Schoolmaster Enoch Wines would describe a Marine as "a sort of ambidextrous animal—half horse, half alligator" whose duties "alternated between those of a sailor and soldier," and it was this fact that caused many contemporaries to believe mistakenly that the Marines were either part of the Navy or the Army rather than their own separate service.

When Congress declared war on Great Britain in 1812, the Marine Corps mustered slightly fewer than 1,300 officers and men out of an authorized strength of 1,869. Despite the government's efforts to increase the Corps' strength, the number of Marines under arms throughout the war remained fairly level. With Marine privates drawing the lowest pay of all three services (six dollars a month) and ineligible to receive the same cash and land bounties given to the military, the inability of Marine recruiters to fill the Corps' rolls is understandable. Lieutenant Colonel Commandant Franklin Wharton divided his Marines among U.S. warships on the Atlantic seaboard, Chauncey's and Perry's fleets on Ontario and Erie, and Navy stations and yards from Portsmouth, New Hampshire, to New Orleans, Louisiana. As a 44-gun frigate, *Constitution* was authorized to carry a complement of sixty Marines: two lieutenants, three sergeants, three corporals, a fifer and drummer, and fifty privates. Although *Constitution*'s Marine detachment was seriously understrength in the summers of 1812 and 1814, the frigate carried close to its full complement for the rest of the war. The commander of *Constitution*'s Marines at the war's commencement was Lieutenant William S. Bush who was killed in the engagement with *Guerriere* and whom Congress awarded a medal posthumously for gallantry in that action. Captain Archibald Henderson, who would later serve as one of the Corps' most distinguished early commandants, commanded Old Ironsides' Marines on its last wartime cruise.

The most important role played by Marines in *Constitution* during the War of 1812 was as sea soldiers. To this end the crew was trained regularly in marksmanship and the manual of arms. When the ship went into action against an enemy vessel, its Marines stationed themselves at various places on the deck (hatchways, gangways, and quarterdeck) and aloft in the ship's fighting tops. Once the frigate closed to pistol range with its opponent, Marine marksmen delivered their fire against enemy targets in the opposing ship. Marines in the tops sometimes worked the small howitzers or swivels mounted there, firing grapeshot at enemy personnel. The most tempting targets, if they could be seen through the smoke, were the officers on the enemy's spar deck. Well-aimed fire from the fighting tops could prove telling in actions fought at close quarters. In *Constitution*'s engagements with *Guerriere* and *Java,* sharpshooters in Old Ironsides managed to disable or kill the captains and several senior officers in both British ships as well as cut down many of the enemy crew. *Constitution*'s steady, well-directed Marine musket fire likewise contributed to the American 44's victory over *Cyane* and *Levant.* Other soldierly duties *Constitution*'s Marines were expected to perform included assisting boarding parties, repelling enemy boarders, serving the ship's guns, and participating in amphibious operations against enemy ships and fortifications.

Constitution's fore and maintops provided a platform from which its Marines could deliver musket fire in battle. Here, *Constitution*'s Marines fire on HMS *Cyane* from the frigate's maintop in the action of 20 February 1815. *William Gilkerson, watercolor, 20th century, reproduced courtesy of the USS Constitution Museum, Boston, Massachusetts*

A second critical role performed by *Constitution*'s Marines was that of shipboard police. It was the Marine's "principle duty," young man-of-war's man Charles Nordhoff would recollect, "to spy out and bring to punishment all offenders against the laws of the vessel." This included those who pilfered ship's stores, smuggled liquor, left the frigate without permission, or behaved insubordinately. To maintain order and discipline in *Constitution,* Marine sentinels kept watch at various stations throughout the frigate including the captain's cabin, the spirit and storerooms, the grog tub and the galley, as well as the quarterdeck, forecastle, and gangways. They kept guard over wrongdoers and stood under arms when all hands were called to witness punishment. The police duties carried out by Marines in Navy ships promoted a hearty contempt on the part of Sailors for these sea soldiers. According to Nordhoff, it was this contempt that inspired the saying "a messmate before a shipmate, a shipmate before a stranger, a stranger before a dog, but a *dog* before a soldier."

In addition to their military and constabulary duties, *Constitution*'s Marines served as a ceremonial guard. Whenever the ship's captain, or high-ranking officers and civil officials visited the ship, a Marine guard was obliged to be on hand to greet them. The Marines were the public face of the ship and as such were expected to appear clean and properly dressed when turned out for ceremonial functions. Marines were required to keep their arms in order and their uniforms clean so that they always presented a smart appearance on such occasions.

Had it been left to *Constitution*'s Marine commander, his detachment's responsibilities would have been restricted to those described above, but Marine enlisted men were expected to share (in a limited degree) in the work of running the ship. Though excused from going aloft, any duty requiring brute strength, lifting, pushing, or hauling became their lot to perform when not on sentinel duty. For example, serving at the capstan when orders were given to raise the anchor or hauling on ropes when the frigate went through particular sailing maneuvers. Marine officers in *Constitution* and throughout the fleet viewed such service as an infringement on their command authority and detrimental to the well-being and morale of the men in their detachments. Disagreements over the charge of Leathernecks aboard ship were at times serious enough to poison wardroom relations between Marine and Navy officers, prompting some of the more hotheaded disputants to settle their arguments with dueling pistols.

Fore Top—
John Day — Mas^r Mate
John Wentworth
William Moore —
Asa Curtis
George Hands —
Allen Mc Donald — Corp^l M
Andrew Chambers — Private
Thomas Phipps — do
John German — do
Peter Wayberg — do

A page from *Constitution*'s 1812 quarter bill listing Marines and Sailors whose battle station was the ship's foretop. *Naval Historical Center photograph*

Marine Corps historian Allan Millett writes that the contributions of Marines in combat during the War of 1812 were problematic. Seamanship and gunnery, he states, not musket fire and boarding actions, won American victories at sea. Yet the Marines in *Constitution* did play a valuable role in the triumphs over *Guerriere, Java, Cyane,* and *Levant,* as did Marines in other American ship and fleet victories. Moreover, Marines gave a good account of themselves in actions ashore such as at the Battle of Bladensburg and in the defense of New Orleans. Through their courage and sacrifice, the Marines of *Constitution* and of the entire Corps shared in the good will and enhanced reputation earned by the Navy during the War of 1812.

CEB

Salt Junk and Ship Bread: Feeding the Men of *Constitution*

One of the most important responsibilities of an American ship commander in the War of 1812 was keeping his crew well supplied with food and water. Ample rations of meat, bread, and drink enabled Sailors to endure the fatigues of shipboard duty; provided the nourishment necessary to fight off illness or recover from injury; and sustained morale in times of danger and difficulty. Provisions were thus integral to a crew's performance, health, and fighting spirit. They were also critical to a ship's operational range, for no Navy vessel could long remain at sea without a well-stowed hold of foodstuffs and water. With the British blockade growing in effectiveness as the war progressed, the Navy had need of resourceful captains who, once at sea, were capable of extending their cruises through the wise and careful management of provisions. The outstanding exemplar of such an officer was David Porter, who cruised in the frigate *Essex* for seventeen months before suffering capture.

By law, the Navy was obliged to provide each serviceman with a daily ration of food. Officers were entitled to receive extra rations (one to eight depending on rank), which they could draw in kind, or convert to cash at a fixed rate per ration ranging from 20 cents in 1812 to 25 cents after 1813. The component parts of this ration were meat (beef and pork), suet (beef fat), bread, flour, vegetables (peas and rice), molasses, vinegar, and spirits (rum or whiskey). These victuals were issued according to a table that defined the type and amount of provision to be served out each day of the week (See p. 142, U.S. Navy Ration.) For example, on Mondays, a Sailor could expect to receive fourteen ounces of bread, one pound of pork, half a pint of peas, and half a pint of distilled spirits as his daily ration. A ship's bill of fare for the remaining six days of the week was equally well-known to any member of its crew.

The cook prepared hot meals for the crew using a wood-fired stove called a galley or "camboose."
Old Ironsides, U.S. Frigate Constitution: An Essay in Sketches

While Sailors dined regularly on salted meat ("salt junk") and hard-baked biscuit (ship bread), they also occasionally received fresh food at mealtimes. In fact, departmental regulations required crews to be "supplied with fresh provisions" whenever convenient. Captains preparing their ships for extended voyages customarily laid in stores of fruits and vegetables and brought livestock aboard to furnish the crew with fresh meat. Visits to foreign ports presented another opportunity for Navy captains to supply their men with fresh foodstuffs. During *Constitution*'s visit to the Brazilian island of Fernando de Noronha in early December 1812, William Bainbridge replenished Old Ironsides' provisions with watermelons, bananas, cocoanuts, fresh fish, and pigs. On occasion American tars satisfied their appetites for fresh meat by hunting or gathering up animals ashore. The men of the frigate *Essex* feasted on fish, crabs, birds, iguanas, and turtles in the Galápagos Islands throughout the spring and summer of 1813. Sailors also supplemented their rations with purchases of tea and sugar from the ship's purser or by pooling their resources to buy fresh food ashore or from bumboats while in port. Officers used the moneys from converted rations to stock their own private pantries with food and drink.

In addition to regular rations of meat, bread, and vegetables, each man aboard ship received a daily allowance of water to cook his food and quench his thirst. The captain was responsible for setting the amount of this allowance. What was considered a proper daily water ration for U.S. Sailors during the War of 1812 is unclear, though one gallon (the allowance first mentioned in Navy Regulations in 1833) is the most likely amount. If circumstances such as depleted, contaminated, or scarce water supplies necessitated conservation measures, captains were authorized to reduce their men's water allowance to two quarts. When it is considered that this two-quart amount included water both for cooking and drinking, and, that on average, the human body requires between two and a half and three quarts of water a day to maintain itself properly, the hardship such a reduction imposed on a ship's company can well be imagined, especially for vessels operating in hot climates. William Bainbridge and Charles Stewart each put their men on a two-quart water allowance during wartime cruises in *Constitution*.

Because water was so important to the Sailor's diet, maintaining adequate supplies of that vital nutrient was a high priority for commanders at sea. Though dependent on weather, the easiest way to replenish a ship's water supply was by collecting rainwater. More often, water was taken in upon a ship's making landfall, at which time parties were sent ashore to fill large casks with water and ferry them back to the ship. Sometimes watering parties unknowingly returned to the ship with water tainted with human or chemical pollutants. The result was usually widespread sickness among the crew. For example, in the fall of 1804, many in *Constitution* were struck down with severe diarrhea after drinking water obtained at Malta that was "empregnated with a limy substance." Likewise, in the fall of 1845, large numbers of crewmen in USS *Columbus* suffered from dysentery after drinking impure rainwater brought aboard while visiting Java.

The purser assumed responsibility for the frigate's provisions after they were stowed on board. This officer conducted weekly inspections of the hold to ensure that the ship's food stores remained wholesome and unspoiled. He kept a record of all rations issued to the crew and of all provisions lost due to spoilage or theft. He also purchased provisions for the ship after it was under way. The purser's steward managed the distribution of each day's rations to the crew. Meat was delivered to the ship's cook the evening before preparation, and placed in a steeping tub filled with fresh water to leach away the salt preservatives. Two petty officers oversaw the division of the meat into equal portions for each mess. The meat was then tagged with each mess's tally marker and placed for cooking in a large copper sunk in the "camboose" or galley. The cook took a sample of the cooked meat to the officer of the deck for tasting before portioning it out to the crew.

To facilitate the serving out and eating of provisions, *Constitution*'s company was formed into messes with rank as the organizing principle. The captain dined alone in his stateroom and had his own servant and cook to prepare his food. Wardroom, steerage, and warrant officers messed in their respective quarters and, like the captain, had their own cooks and servants to assist with mealtime preparations. The rest of the crew took their meals on the frigate's berth and gun decks grouped in messes of eight men. Each mess appointed one of its members (the "caterer" for officers, the "mess cook" for enlisted) to be in charge of meal preparation and cleanup, a duty that, among the enlisted, was rotated on a weekly basis. The senior crewman in each enlisted mess presided as its head and was responsible for maintaining good order among messmates.

If emergency or circumstance did not interfere, *Constitution*'s crew ate three meals a day: breakfast at 8 a.m., dinner at noon, and supper at 4 p.m. One hour was allotted for meals with one half the crew (the offgoing watch) eating first, followed by the other half (the ongoing watch). Officer's messes sat down to their meals in order of ascending seniority after the men were fed. The crew ate picnic-style on a tarred, canvas cloth spread between the guns, using assorted mess ware such as tin cups, basins, forks, knives, and spoons. Dinner, the one hot meal of the day, was served out at the galley and carried back by the mess cooks in tubs called "kids." A number of Sailor's dishes had colorful names such as lobscouse (corned beef hash), burgoo (oatmeal porridge) and duff (steamed suet-pudding with raisins). Breakfast and supper usually consisted of cold leftovers. A drum roll following dinner and supper summoned the crew to receive their spirit ration at the gun deck mainmast. This ration, sometimes issued raw, was usually served out as a one-to-one mixture of water and rum or whiskey ("grog") to the messes by turn. Each man had to down his half-pint measure of grog on the spot. Commissioned, warrant, and sometimes petty officers were allowed to drink their spirit rations at their mess tables.

Given the primitive state of food preservation in the early nineteenth century, the types of provisions (pickled, salted, and dried) that made up the Navy ration were the best that could be had under the circumstances. And if they did not provide the tastiest fare for *Constitution*'s

The Supper. p. 45.

Mealtimes offered Sailors a welcome break from the day's fatiguing labors. *Naval Historical Center photograph*

Sailors, they were issued in sufficient quantity to sustain them through their labors. One modern study of the nineteenth-century naval diet estimates the American Sailor's daily ration to have contained 4,240 calories, well above the 2,500 to 3,000 calories nutritionists state are required by today's active male. Where the food served in Navy ships fell short was in vitamin content, particularly Vitamins A and C. Deficiencies in the former can lead to night blindness while severe shortages of the latter result in scurvy. Charles Stewart terminated his first cruise in *Constitution* prematurely, in part because of an outbreak of scurvy in Old Ironsides. He attributed the disease's appearance to the ship's lack of fresh provisions.

Shortages of provisions aboard ship, when they did occur, could produce profound suffering, as when the crew of *Essex* was reduced to eating rats and ship's pets when the frigate's stores fell to perilous levels. They also threatened morale. On the outward-bound voyage of *Constitution* in the late fall of 1812, William Bainbridge found himself confronted by a hungry and mutinous gathering of Sailors complaining of their short allowances of food and water. Bainbridge's resolute manner and words of explanation defused the situation and the disgruntled Constitutions returned to their duty. John Sinclair, captain of the privateer *General Armstrong,* did not fare as well as Bainbridge in facing down a hungry and thirsty crew. In March of 1813, in waters north of the Cape Verde Islands, Sinclair's crew, frustrated with their commander's ineptitude, and angry at serving on short rations of food and water, rose up and seized control of the privateer. Episodes such as these illustrate the risks inherent in tampering with a Sailor's ration.

CEB

Colt and Cat: Naval Discipline in the War of 1812

Three sets of complementary regulations governed the conduct of Navy personnel during the War of 1812. Congress enacted the first and most important of these on 1 June 1800 in "An Act for the Better Government of the Navy of the United States." (See pp. 114-19 for the text of this act.) This statute, also known as the Articles of War, defined the standards of behavior for officers and enlisted men, identified violations of Navy law and their corresponding punishments, described the administrative duties of commanding officers, specified the rules for Navy courts, and established the formula for distributing prize moneys. Captains of all public vessels were required to read the Articles of War to their assembled crews once a month. The Articles, with numerous revisions over time, formed the heart of the Navy's disciplinary code until 1950, when Congress enacted the Uniform Code of Military Justice (UCMJ). The second body of rules regulating naval personnel in 1812 was the Navy Department's own regulations. (See pp. 120–29 for the text of these regulations.) These instructions, issued by authority of the president in 1802, outlined the duties and responsibilities of naval officers and certain classes of petty officers. The final set of rules guiding Sailors' behavior were those issued by individual captains to regulate personnel, activities, and routine aboard their ships.

The Articles of War provided for the punishment of several different categories of misconduct: actions that were destructive of good morals (drunkenness, swearing, theft); those that threatened the ship's safety (negligence, desertion); those that undermined the ship in battle (cowardice, disaffection); and, those injurious to authority (mutiny, insubordination). Men found guilty of violating one or more of the Articles were subject to a range of punishments including fines, confinement in irons, suspension or dismissal from service, and flogging. Thirteen articles carried the ultimate sanction of death. The Articles also permitted naval authorities to punish wrongdoers according to the "laws and customs" of the sea. Reductions in rank, stoppages of rations, and the wearing of badges of disgrace were just some of the penalties meted out to those convicted under this proviso. It should be noted that Marines came under governance of the Navy's Articles only when they served on board ships. When stationed ashore, the Army's Articles of War regulated their discipline.

The chief tool Navy captains and their subordinates employed to exercise disciplinary control over enlisted men was corporal punishment. The Articles of War permitted the infliction of up to a dozen lashes on Sailors guilty of misdemeanors or of offenses not serious enough to warrant a court-martial. This type of punishment was applied in one of two ways. The first was the use of a short length of rope known as a "colt" to "start" or strike misbehaving seamen. Moses Smith, a Sailor who served in *Constitution* under Isaac Hull, recalled being chastised in this manner when he and thirty of his fellow topmen received blows from a "rope's end well laid on" for displeasing the officer of the deck. In addition to this on-the-spot form of justice, captains had the authority to confine seamen for brief periods or flog them up to twelve lashes. *Constitution's* wartime logbooks detail nearly two-dozen examples of this type of punishment. For instance, on 12 August 1812, *Constitution*'s log recorded that John W. Smith and John Smith received "one dozen lashes each," the former for drunkenness and the latter "for insolence to his Officer on duty." Nearly three months later Private Anthony Reaves was confined for a week and given a dozen lashes for neglect of duty.

Ship commanders punished more grievous breaches of naval law through courts-martial. In the pre-1815 Navy, the two crimes tried most often before these tribunals were desertion and mutinous or seditious conduct. The authority to convene courts-martial resided in the president, the secretary of the navy, and commodores at sea or on foreign stations. Therefore, commanding officers first had to forward a request for trial to proper officials, usually the Navy secretary, before proceeding against wrongdoers. Once the secretary authorized a court, he issued orders appointing a board of naval officers to hear the case and a judge advocate (sometimes a civilian) to serve as prosecutor and court recorder. Unlike today's system of military law under the UCMJ, accused Sailors in 1812 did not enjoy the right to legal representation at their hearings. Court verdicts were subject to review by commanding officers and by the secretary of the navy, who had the power to direct courts to reconsider their findings. William Bainbridge did just this when he ordered the court that had acquitted Private John Pershaw of theft to reassess its verdict. After reviewing its decision, the court declared Pershaw guilty and sentenced him "to have his head shaved and to be dismissed [from] the service with every mark of disgrace." The severest penalty naval courts could hand down in non-capital crimes (that is cases not punishable by death) was one hundred lashes. Capital offenses required a two-thirds majority of the court to award the death penalty, which in turn required the president's approval before being carried into execution. Sailors convicted of capital crimes but not sentenced to death were punished with significantly higher totals of lashes. Seaman Jean Baptiste, for example, received two hundred lashes for deserting the Charlestown Navy Yard in October 1814 to join the privateer *Leo*.

The spectacle of a flogging was meant to inspire terror among a ship's crew thereby deterring misconduct and ensuring obedience to naval law. The shrill sound of the boatswain's pipe, followed by the call, "All hands to witness punishment ahoy," signaled the commencement of the flogging ritual. With the ship's officers gathered on the starboard side of the quarterdeck, and an armed Marine guard drawn up on the larboard side, the master at arms escorted the prisoner to the gangway, where his shirt was stripped off and his hands and feet secured between the hammock rail and grating on deck. The crew assembled around the gangway while the ship's surgeon positioned himself nearby to assess the prisoner's condition during punishment. On signal, the boatswain's mate removed from a canvas bag a thick rope with nine knotted cords attached to it known as a cat-of-nine-tails (hence the expression "letting the cat out of the bag"). At the captain's command, "Boatswain's mate, do your duty," the flogging began. The master at arms counted out each stroke of the cat across the prisoner's bare back until the punishment was complete. One can imagine the chilling effect such scenes had on the minds of a ship's company, especially if it was repeated with multiple prisoners. Charles Nordhoff, who served as a boy in the ship of the line *Columbus* in the 1840s, was so sickened the first time he witnessed a flogging that he covered his eyes to avoid viewing the nineteen additional whippings that followed. "Never more did I *see* a man flogged," the young Nordhoff wrote.

Although flogging was a common feature of life in the early sailing Navy, it would be a mistake to conclude that all naval officers favored the lash for maintaining order over their men. A number of captains, such as Isaac Hull, found the practice of flogging abhorrent, resorting instead to lesser punishments, whenever possible, to correct misbehavior. Hull's humane views on discipline and his concern for the well-being of the common Sailor, endeared him to the enlisted ranks serving under him. As a veteran of Hull's Mediterranean flagship *Ohio* would reminisce in 1841: "Every man of his crew honor and admire him, for his love of justice and clemency, and a devotion he always manifested in behalf of their happiness and welfare." The Constitutions' affection for Hull ran so deep that when William Bainbridge, an officer reputed to be a

Naval authorities believed that religious instruction contributed to good order and discipline in the fleet. Larger vessels carried chaplains who ministered to the men's spiritual needs. *Naval Historical Center photograph*

Contemporary sketch of a flogging administered aboard USS *Porpoise*. Congress abolished this instrument of naval discipline in 1850. *Naval Historical Center photograph*

stern disciplinarian, assumed command of the frigate in September 1812, it occasioned a near-mutiny among the crew, who begged Hull to remain as their captain.

Each Navy vessel had a number of enlisted men who assisted officers in the enforcement of shipboard discipline. Chief among these was the master at arms. Nicknamed "Jemmy Legs" by his shipmates, the master at arms served as the ship's head of police. He had charge over all prisoners, assisted at floggings, searched the ship for contraband, punished the slovenly and lazy, and kept a running list of all crewmen guilty of misdemeanors (the blacklist). Author Herman Melville wrote that masters at arms were such hated figures aboard ship that on "dark nights . . . [they] keep themselves in readiness to dodge forty-two pound balls, dropped down the hatchways near them." The master at arms was assisted in his constabulary duties by the ship's corporal. Additional personnel who helped discipline the crew and maintain order aboard ship included the boatswain, his mates, and Marines.

Surviving documents reveal that eight of Old Ironsides' enlisted men (including two Marines) were court-martialed during the War of 1812, seven while the frigate was homeported at Boston in 1813 and 1814, and one while it was at sea in November 1812. The charges in these cases included petty theft, assault, mutinous conduct, and desertion. Of the eight accused Sailors, one was acquitted, another was cashiered, and the remaining six were awarded floggings of between twelve and one hundred lashes. Two of this last group were sentenced to the additional penalty of wearing a sign labeled "deserter" for one month, while a third was reduced in rate from sailmaker's mate to ordinary seaman.

Officers could prove as disruptive to a ship's discipline as any ill-behaved seaman and the officers of *Constitution* proved no exception. Midshipman James W. Delany was one officer who proved especially exasperating to his superiors. In late August 1813, the young Delany was arrested after appearing drunk in the streets of Salem, Massachusetts. Though ultimately acquitted of the charge, the young warrant officer found himself again in hot water that December for permitting five Sailors to desert in a cutter. For this action Charles Stewart ordered Delany out of the ship. Despite William Bainbridge's recommendation that Delany be dismissed from the Navy, the Secretary declined to punish the midshipman. Other examples of officer misconduct in *Constitution* include Midshipman Joseph Cross, court-martialed for insubordination; Midshipman Shubael Pratt, set ashore for permitting the crew of a guard boat to plunder a local coasting vessel; and Midshipman Jott S. Paine, dismissed for assaulting fellow Midshipman Zachariah W. Nixon. The most noteworthy wartime disciplinary case among the ship's officers involved Charles Stewart, who was summoned before a court of inquiry to explain why he brought his first cruise in *Constitution* to a premature conclusion in early April 1814. The chastened Stewart was found guilty of an error in judgment but allowed to retain command of Old Ironsides.

The majority of *Constitution*'s officers performed their duty in exemplary fashion. But because this class of personnel was exempt from corporal punishment, the frigate's commanders found it difficult to discipline some of their more refractory subordinates. Admonishment, confinement, and suspension from duty were useful tools to recall officers to their duty. When these methods failed, only the vehicle of courts-martial remained.

Corporal punishment was the keystone of the Navy's disciplinary edifice during the War of 1812, reflecting a belief that the Sailors who served in the nation's fleet were a tough, hard-bitten class of men who could only be governed by brute force. Until this view of the serving Sailor changed, the Navy's leadership remained wedded to the lash as an instrument of discipline. The Navy took the first momentous step toward establishing a more humane disciplinary philosophy when Congress abolished flogging in 1850. This action inaugurated a sea change in naval law and looked to a future when ideals of honor, courage, and commitment would spur Sailors to duty rather than fear and violence.

CEB

The Great Chase

Isaac Hull had commanded USS *Constitution* for two years when America declared war on Great Britain in June 1812. Besides patrolling the coast during this time, Captain Hull had overseen the frigate's several overhauls to improve its sailing qualities. The previous summer, the Navy Department had ordered *Constitution* to ferry several diplomats around Europe and to deliver a debt payment to Holland. Several encounters with British squadrons created tense moments but ended without incident and the ship returned to the Chesapeake Bay in February 1812.

Finding on his return from Europe that America was drifting toward war, Hull spent the next months preparing *Constitution* for that eventuality. Newly rigged and outfitted, *Constitution* left the Washington Navy Yard in early June 1812 and was en route to Annapolis on 18 June when Congress declared war. Navy Secretary Paul Hamilton advocated a strategy to protect the few naval resources held by the Americans. Ordering his captains to cruise in squadrons, Hamilton directed Hull to join John Rodgers in New York. Meanwhile, a British squadron had sortied from Halifax to intercept Rodgers's squadron.

On 16 July at 2 p.m., Hull sighted four sails. In pursuit by 4 p.m. *Constitution* saw a ship standing towards it and possibly others near shore. After approaching within six or eight miles of the ships and not receiving a response to the private signal, Hull at 11:00 p.m. determined that the strange sails belonged to the enemy. Hull, however, decided to avoid action until daylight in order to spare his untrained crew from the confusion of a nighttime engagement. No wind at sunrise on 17 July becalmed *Constitution*, rendering the ship unmanageable. Daylight also brought not one but half a dozen vessels in sight. Desperately pursuing all means of escape, Hull ordered his crew to pump water overboard in order to lighten the ship and to wet all the sails, closing the texture of the canvas, thus eking out every small advantage of breeze possible. *Constitution*'s reputation as a poor sailer convinced the crew that the situation was hopeless. Keeping out of firing range to prevent being

Isaac Hull resorted to several stratagems (in this scene, kedging) in his desperate attempt to outsail and escape a pursuing British squadron during the first month of the War of 1812. *Navy Art Collection, Naval Historical Center*

overtaken, Hull resorted to kedging his ship. Kedging is a slow, laborious way to move a vessel hindered by a confined area or the lack of wind. Seamen rowed a longboat containing a small anchor out the distance of a cable, dropped it, and manually hauled in the ship using the capstan. The Americans began kedging at 7:00 a.m. and the British followed suit once they found themselves becalmed. Having more men and boats, the British concentrated on towing those ships closest to *Constitution* and thus held a decided advantage in closing the distance to within firing range.

Besides lowering two cutters to tow the ship, the crew hoisted a 24-pounder from the gun deck to the spar deck and positioned it at the taffrail along with an 18-pounder to serve as stern chasers. An additional two 24-pounders were pointed out the captain's cabin windows, also to act as stern chasers. At 9:00 a.m., HMS *Belvidera* fired its bow guns and *Guerriere* delivered a broadside. The shots fell short, whereas the shot from *Constitution*'s stern chasers reportedly struck *Belvidera.* All continued kedging but neither side gained ground. To lighten the ship, Hull pumped over 2,300 gallons of drinking water overboard. To improve their chances of overtaking *Constitution*, the British decided to put all their efforts in kedging one ship, H.M. frigate *Shannon.* Throughout the day on the 17th, the crews alternated between towing and resting on what Tyrone Martin, one of *Constitution*'s twentieth-century commanders, called a "diabolical treadmill." Hull had little choice but to risk this maneuver or chance being trapped against the coast. The crew's exertions throughout the night resulted in almost a three-mile lead.

The vigilant Hull skillfully used a light breeze at sunrise to attempt to break away from his pursuers. In doing so, however, he chanced passing close by HMS *Aeolus*, whose captain's decision not to fire on the Americans (probably because the resulting concussion would becalm his ship) widened *Constitution*'s lead. Perhaps a defining moment for the ship came at 9:00 a.m. on the 18th when *Constitution*, having outsailed the enemy, took the time to help an American merchantman escape. Hull scared off his countrymen by hoisting British colors. By 4:00 in the afternoon, six miles separated the foes, but Hull saw an approaching squall as an opportunity to increase his advantage. About 6:45 he shortened sail in preparation for the bad weather, and the British followed suit. The latter were caught by surprise, however, when Hull's crew, hidden by the downpour, sheeted home, setting topgallants and the main topsail staysail. *Constitution* sped away attaining eleven knots. The British kept up the chase for another twenty-four hours, but finally abandoned the pursuit by eight o'clock on Sunday morning, 19 July.

For almost three days the Sailors of *Constitution* endured many anxious moments. Fear mounted when the enemy approached within firing range, but sighs of relief accompanied the shots falling short. The Americans avoided capture through a combination of strong leadership, teamwork, and the tactical mistakes of their adversaries. Captain Hull remained on the deck throughout the long ordeal, anticipating and acting—eventually leaving the enemy in his wake. While the enemy loomed astern, *Constitution*'s crew persevered because their captain would not accept defeat even when there was only the slightest chance for success. By kedging when becalmed and setting sails quickly when a puff of wind developed, the Americans prevailed through superior seamanship. The cocky British squandered their advantages through miscalculations. Hull's immediate subordinate, Lieutenant Charles Morris, surmised that the British wanted to take an undamaged ship as a victorious trophy in the first month of the war and just assumed that *Constitution* would surrender without a fight in the face of overwhelming odds. Instead, Morris noted that if the British had concentrated all their effort from the start in bringing one ship within firing range, they might have succeeded. The British held the numerical advantage, but it was Hull's fashioning a green crew into a team that won the day. From this ordeal, Hull's most recent biographer, Linda Maloney, concludes, "the crew, the ship, and the commander had become one.... The 'Constitutions' were now prepared to go to war." For related documents, see pp. 94–97.

CFH

Constitution versus *Guerriere*

After the hairbreadth escape from the British squadron, *Constitution* stood for Boston and arrived there on 26 July 1812. That port was only a temporary refuge for replenishing supplies, as Captain Hull, eager to evade any blockaders and anxious to prey on the enemy's commerce, sailed within a week, taking advantage of the first favorable wind from the southwest. Seizing the initiative and not waiting for orders that might have kept him in port or taken away his command of *Constitution*, Hull boldly charted a northern course against British shipping in the Halifax–Gulf of Saint Lawrence area. While at Boston, Hull had purchased charts for places as far afield as Brazil and Africa, signaling his willingness to seek out the enemy in all parts of the Atlantic. Weighing *Constitution*'s anchor and setting an easterly course on 2 August, Hull sailed with full confidence in his men and his ship.

While making his way along the coasts of Nova Scotia and Newfoundland, Hull exercised his men in

In a short, thirty-minute engagement, *Constitution* reduced HMS *Guerriere* to a shattered, dismasted hulk, unfit even to be towed into port. *U.S. Naval Academy Museum*

gunnery to hone their skills and relieve the tedium of days without sightings of enemy vessels. The first two weeks of August found the American frigate with only a few paltry catches—some British merchant brigs of little value and a recaptured American vessel. The intelligence about the whereabouts of enemy warships that he garnered from every encounter, however, proved more worthwhile, and Hull stood to the south in search of a reported British frigate. The Americans first spied an unidentified warship at 2:00 p.m. on 19 August. The pursuit commenced. Within an hour and a half Hull found a frigate, one that also eagerly wanted a duel.

Hull took in some of his frigate's sails and hauled up some courses, thus slowing his approach toward the enemy. The maneuvering between the two ships continued until 3:45 when *Guerriere*'s captain, James R. Dacres, backed his frigate's main topsail. Meanwhile, Hull reduced sail to "fighting canvas" in order to minimize damage to sails and rigging. Each side set its colors, commencing first with the British hoisting three ensigns, followed by the Americans with four. Once ready to close with his antagonist, Hull shaped a course for *Constitution* that exposed his ship to raking fire from *Guerriere*'s broadsides. To minimize damage, Hull sailed toward *Guerriere* by yawing his ship (turning its head from side to side) and by steering toward the enemy's starboard quarter. Anticipating that *Constitution* would soon be within raking range, Dacres fired first with a starboard broadside and then instantly wore ship (making a half circle) so that he could fire his port battery. Both captains repeated the same maneuvers—*Guerriere* wearing and *Constitution* yawing—several times, progressively getting closer to each other. Finally, Dacres, ready to close with the enemy, adopted a course nearly before the wind. *Constitution* was now at a disadvantage because the enemy's stern guns proved more effective than the American's bow guns. Hull remedied this situation by setting the main topgallant, which increased his speed and permitted him to draw up to his enemy at 6:00 p.m. Now the jockeying for position ended and the battle began.

The next twenty minutes of side by side dueling produced the following result: *Guerriere* suffered an early crippling blow when its mizzenmast fell over the starboard side and, dragging in the water, turned the ship in that direction. Meanwhile, *Constitution* suffered damage to some braces, halyards, and the fore royal truck. The American frigate earned its nickname, Old Ironsides, during the cannonading when a British shot bounced harmlessly off the hull, prompting an American tar to yell, "Huzza! Her sides are made of iron." While maneuvering, the two ships collided, and *Guerriere*'s bowsprit damaged the gig in the American's stern. Confusion abounded on board both ships but the Americans fired off two broadsides while the British replied with only two feeble shots. Taking advantage of their locked positions, each side seized the opportunity to board. Fierce musketry from both antagonists and high seas stymied the assaults and then *Constitution*'s forward motion separated the two ships—ending further boarding attempts. As the two pulled apart, *Guerriere*'s weakened foremast toppled over, causing the mainmast to fall as well. With its foe dismasted, *Constitution* hauled off close by to repair some rigging before returning to action. Captain Dacres, however, recognized the futility of further fighting and, just before 7 p.m., fired a gun to leeward, the customary sign of surrender.

The extent of the damage suffered by *Guerriere* prevented the Americans from returning home with their prize, as its hold was taking on water faster than the pumps could discharge it. Having to set on fire the unsalvageable British frigate mandated *Constitution*'s early return from its cruise because the American frigate had to accommodate so many prisoners. The tumultuous reception Boston showered on the returning heroes contrasted sharply with British incredulity upon learning of *Guerriere*'s defeat.

The defeat of British frigate *Guerriere* on 19 August 1812 garnered *Constitution*'s captain, Isaac Hull, lasting fame in the annals of naval history. *U.S. Naval Academy Museum*

This American victory unnerved the British. Not mincing words, the *London Times* opined, "How important the first triumph is in giving a tone and character to the war. Never before in the history of the world did an English frigate strike to an American." While this defeat deeply affected the British psyche, it was not that nation's first capitulation of a frigate to an American ship, as H.M. frigate *Fox* struck to the Continental Navy frigates *Boston* and *Hancock* during the War of Independence.

Ever since the battle, historians have debated whether *Constitution*'s superiority in force precluded any chance of success for the British. Many in England, including Captain Dacres and his court-martial board, attributed the defeat to the disparity in men, guns, and weight of broadside, as well as to the bad luck of an "accidental" dismasting. Indeed, the Americans outnumbered the British in men, 450 to 275-300, and in guns, 55 to 49. The proportional difference in broadside weights was 3 to 2. *Guerriere*'s fate was probably sealed early on when its mizzenmast fell over its starboard quarter and shortly thereafter the foremast crashed down carrying the mainmast with it. The British ascribed the loss of the masts to their "defective state," whereas the Americans credited their better gunnery.

The disparity between the force levels of both ships cannot explain the disproportionate casualties incurred—*Constitution* suffered fourteen and *Guerriere* seventy-nine killed and wounded. Many historians conclude that a combination of force and execution determined the outcome, with superior seamanship and gunnery giving the Americans the advantage. *Constitution* was a disciplined man-of-war. Hull cautiously maneuvered his ship for a close action, unlike Dacres who impatiently fired his long guns in hopes of disabling the Americans from a distance. Even when the ships were almost alongside, Hull waited while *Guerriere*'s gun crew fired on the up roll of the waves, sending the shot high into the rigging with the intention of damaging *Constitution*'s sailing ability. The Americans, however, were taught to fire on the down roll, striking the hull and wreaking havoc throughout the British ship. In a public statement issued soon after the engagement, Hull credited his crew for their "gallantry" and "good conduct" in so destroying *Guerriere* that it could not be towed into port. A determined crew combined with good fortune won *Constitution* its first victory in the War of 1812. For related documents, see pp. 98–102.

CFH

Constitution versus *Java*

Constitution returned to Boston Harbor on 30 August 1812 after destroying HMS *Guerriere*. Coincidentally, John Rodgers's squadron of *President*, *United States*, *Hornet*, and *Argus* anchored the following day off Castle Island, after a disappointing two-month, commerce-raiding cruise. *Constitution*'s commander, Isaac Hull, citing pressing family responsibilities, exchanged positions with Captain William Bainbridge, commandant of the Charlestown Navy Yard. The yard employees worked feverishly in September to refit and resupply the fleet; meanwhile, the Navy Department formulated its strategy for the second deployment.

The immediate goals after the declaration of war were to protect American merchant ships returning to homeport and to reap some captures during the early stages of the conflict from an unprepared enemy. By the fall of 1812, the British recognized that their diplomatic overtures had not swayed the Americans. Preoccupied with the war in Europe, the Admiralty decided to concentrate on convoying its commercial fleet and blockading American ports, rather than commit the greater forces required for major operations. Recognizing the enemy's commercial vulnerability, Secretary of the Navy Paul Hamilton decided to attack British trade routes from the West Indies and South America to England by dividing American naval forces into three detachments that would fan out across the Atlantic on a southeastwardly course toward Africa and Brazil. Two squadrons sortied from Boston on 8 October. Rodgers with frigates *President* and *Congress* sailed toward the Canary Islands, while frigate *United States* and brig *Argus* under Stephen Decatur's flag diverged more to the south. Outfitting hitches delayed the third squadron, composed of Bainbridge's *Constitution* and Master Commandant James Lawrence's sloop of war *Hornet*, from clearing President Roads in Boston Harbor until 27 October. David Porter's frigate *Essex*, also assigned to Bainbridge, never rendezvoused with the others but sought its fame against the British whaling fleet in the Pacific.

Secretary Hamilton's general orders "to annoy the enemy and to afford protection to our commerce" provided Bainbridge with plenty of leeway to formulate his own plans. He decided to take the southernmost sweep of the Atlantic, heading to the Cape Verde Islands first and eventually into the South Atlantic. While *Constitution* was in fighting trim from its recent overhaul, its new commander demanded the crew's obedience and respect. They cheered their departing captain as they openly criticized their new commander, whose reputation for bad luck, poor judgment, and rigid discipline was known throughout the fleet. A number of the crew were arrested for mutinous language on this occasion and others deserted rather than serve under Bainbridge. All commanders seek the glory of naval victory, but a victorious cruise for Bainbridge would erase his past failures. Recognizing the value of molding his men into automatons who could perform their tasks unthinkingly during the worst moments of battle, Bainbridge adopted a regimen of exercising his men in seamanship and gunnery. He worked diligently to ensure success.

Stern disciplinarian William Bainbridge succeeded popular Isaac Hull as *Constitution*'s second wartime commander. *U.S. Naval Academy Museum*

After stopping at several of the prearranged rendezvous and finding no sign of *Essex*, *Constitution* and *Hornet* proceeded to Salvador (Bahia, Brazil), reaching that port by mid-December. Discovering HMS *Bonne Citoyenne* (practically equal in men and guns with *Hornet*) in that harbor, Lawrence challenged the enemy sloop of war to single-ship combat. The British captain declined, probably because his ship had $1,600,000 in specie in its hold to protect. Leaving *Hornet* to blockade the valuable sloop, Bainbridge set out along the Brazilian coast in search of merchant vessels. At 9:00 a.m. on 29 December, *Constitution* sighted two sail about thirty miles off Bahia. The smaller vessel headed toward port, but the larger one, seeing *Constitution* in the south-southeast, shaped its course toward the American warship. Concerned at first that the strange sail was a ship of the line, Bainbridge headed eastward to avoid being pinned against the coast. By early afternoon, after the customary show of colors and signals and noting the other ship's swiftness, the Americans knew that an enemy frigate was in pursuit, as *Constitution* could outsail a liner in those conditions. H.M. frigate *Java*, 38, commanded by Captain Henry Lambert, had diverted to Bahia for water while in transit to India from England. As *Java* was the faster of the two frigates, Lambert could have escaped easily, but decided to engage his foe. *Java*'s speed allowed it to take the weather gauge (a windward position), leaving *Constitution* to struggle against a northeasterly wind.

The two foes later reported different times that certain events occurred during the battle, but both sides agreed that the engagement began about two in the afternoon. With both ships sailing on a southeasterly course, the Americans fired first into *Java*'s rigging, taking advantage of the longer range of *Constitution*'s guns and hoping to slow the British frigate's pace. Bainbridge's strategy of crippling his opponent from afar, however, failed. *Java* waited until within range to open fire. *Constitution* responded in kind and a general engagement followed. *Java*'s first salvo damaged *Constitution*'s spars and rigging and wounded its captain in the thigh. Concerned that the speedier Briton would cross his bow and rake his ship, Bainbridge adroitly fired a broadside and wore around. *Java* mimicked the maneuver and both ships headed westward with the Briton to starboard of the American. Again *Java* tried to sail ahead and again Bainbridge took the initiative, loosing a broadside and wearing ship in the smoke—all to preempt being raked. Now both ships had returned to a southeasterly course but instead of repeating the previous scenarios, Lambert wore his ship just before pulling alongside Old Ironsides and raked its stern with devastating effect—shattering the wheel and wounding Bainbridge a second time in the thigh. Refusing treatment below, Bainbridge remained in command, physically supported by a midshipman, and barked out orders to rig the steering system using tackles attached to the tiller.

Unaware that *Constitution*'s steering problems prevented the Americans from following its course, *Java* thought its foe had disengaged. Lambert jumped at the chance to dispatch *Constitution* and tacked in pursuit, raking Old Ironsides' stern with a broadside. In the meantime, while Bainbridge's wound restricted his movement, it did not impede his mental agility. Recognizing that he had to unleash a crippling blow before his opponent did, *Constitution*'s captain took a bold gamble. Choosing to risk another rake, Bain-

William Bainbridge's expert ship handling, coupled with a disciplined crew, secured *Constitution*'s victory over H.M. frigate *Java*, after a three-hour slug-fest off the coast of Brazil on 29 December 1812. *Naval Historical Center photograph*

bridge set his fore and main courses to close quickly on *Java* and inflict the greatest damage possible with his short-range carronades. The subsequent broadside shot away the enemy's bowsprit cap, jib boom, and headsails. The loss of the headsails finally curtailed *Java*'s maneuverability, leaving the British frigate vulnerable to what proved to be a crushing rake from the Americans. A desperate Lambert realized that an unmanageable ship left him only one option—boarding. His attempt to run down on Old Ironsides failed, however, when *Java*'s bowsprit became entangled in *Constitution*'s mizzen rigging just long enough for a full American broadside and musket shot from the Marines in the tops to wreak havoc on the British. This action severed *Java*'s main topmast and mortally wounded Captain Lambert. Unable to maneuver, *Java* lay helpless to resist a double raking of first the bow and then the stern. American shot toppled the mizzenmast and the remaining section of the foremast. *Constitution* disengaged, effecting repairs a short distance away and *Java*'s crew, now commanded by First Lieutenant Henry Ducie Chads, worked furiously but unsuccessfully to rig sails to the only remaining yard—the lower mainmast. Having to drop that unstable yard to protect *Java*'s deck left the British frigate mastless. *Constitution* returned from its short respite, positioning itself to rake *Java* across its bow. The helpless Briton hauled down its ensign at 5:25 p.m. (American report) or 5:50 p.m. (British report).

No single factor decides the result of a naval engagement. The physical attributes of a ship, the number, type, and weight of guns, and the number and proficiency of the men all determine a battle's outcome. *Constitution*'s thick scantling and stout masts stood it well during its engagement with *Java*. The American frigate did not sustain the crippling damage to its hull and masts that the British frigate suffered. The 44-gun American frigate carried more guns (55-47) of heavier weight (thirty 24-pounder long guns, twenty-four 32-pounder carronades, and one 18-pounder bow-chaser) than its 38-gun adversary (twenty-eight 18-pounder and two 9-pounder long guns, and sixteen 32-pounder and one 18-pounder carronades). *Constitution* also benefited from a complement of 480 to either 373 or 426 men on *Java*. (The British numbers vary and include about one hundred army and navy personnel being transported in *Java* to India.) The wrecked state of *Java* compared to *Constitution*'s was reflected in the disparity in casualties—4 or 5 to 1.

While more men are usually advantageous, their quality is also important. Captain Lambert complained even before he left England that his crew was a motley group of landsmen, obtained from press gangs and prisons. The Admiralty refused to send Lambert better recruits, advising him to use the long voyage to train his men. *Constitution*'s log reveals frequent drills of the ship's company, whereas *Java*'s men reportedly practiced at the great guns only the day before the battle. The great disparity in the ratio of casualties suffered (four or five British to one American) reflected the poor gunnery skills of *Java*'s untrained tars. A disciplined, well-trained crew can better endure a long fight such as this one. *Java*'s gunnery was most telling early on in the engagement, but gradually its accuracy waned. The first British broadside greatly damaged *Constitution*'s spars and rigging and could have caused the loss of a crucial headsail if Seaman Asa Curtis had not slid down the foretopgallant stay and rebent a halyard. A well-run ship survives because of such alert professionalism.

Skilled leadership is the cornerstone of any successful engagement. Bainbridge adroitly maneuvered Old Ironsides, setting the pace of the battle from the beginning and deftly parrying every move from *Java*. All authorities praise Captain Lambert's tactical abilities but Captain Bainbridge's aggressiveness, backed by a superbly designed ship and a disciplined crew, determined the outcome. With one stroke he hazarded all, positioning his ship for a crushing rake, and gambling, successfully, that the enemy was unprepared to take advantage of *Constitution*'s momentary vulnerability. Luck also plays a role in combat. William Bainbridge, arguably one of the unluckiest naval officers in the early Navy, made his own luck by using his ship and men most efficiently. For related documents, see pp. 102–6.

CFH

Charles Stewart and the Capture of HMS *Cyane* and *Levant*

In engaging two British warships simultaneously and capturing both, taking many fewer casualties than those he inflicted on the enemy, Captain Charles Stewart demonstrated the extraordinary mastery of ships under sail attained by the most skillful of the officers of the United States Navy.

In its first cruise under Stewart's command, *Constitution* sailed from Boston in December 1813 and made for the West Indies, where it preyed on British merchant shipping and took and burned H.M. schooner *Pictou*. On 3 April 1814, while *Constitution* was returning to Boston to replace a cracked mainmast and to restore the health of the crew, in which scurvy was making an appearance, two 38-gun British frigates, *Junon* and *Tenedos,* spotted the American warship off Massachusetts and chased it into Marblehead. Later that month the American frigate

Through skillful seamanship, *Constitution*'s Charles Stewart outmaneuvered and captured two smaller British men-of-war, 24-gun *Cyane* and 18-gun *Levant* on 20 February 1815 off Madeira. *Navy Art Collection, Naval Historical Center*

managed to shift anchorage to Boston. Unable to evade the now very strong blockading force off the harbor, Old Ironsides remained at Boston for nearly nine months.

Winter weather having finally driven the British squadron off the blockade, *Constitution* put to sea on 17 December 1814 and sailed into the mid-Atlantic. Examining neutral ships off the coast of Portugal in early February 1815, Stewart learned of news of a treaty of peace having been signed by negotiators at Ghent, Belgium. As these were unconfirmed reports, and as in any case the treaty would need to be ratified before it took effect, Stewart continued his cruise. On 16 February *Constitution* captured the British merchant ship *Susanna,* removing from it two large cats, perhaps jaguars, before dispatching the prize for New York.

On the afternoon of 20 February 1815, *Constitution* spotted a sail while cruising east of Madeira and gave chase, sniping at the fleeing vessel with 18-pounder bow chasers. Having to reduce sail in order to replace the main royal mast that carried away, Stewart was unable to prevent a rendezvous between what turned out to be the 24-gun ship of war *Cyane,* Captain Gordon Thomas Falcon, and the 18-gun ship *Levant,* Captain George Douglas. Although the two British commanders attempted to postpone action until dark, when they might escape entirely, Stewart descended on them rapidly, and shortly before 6 p.m. they went to fighting sails—topsails and jibs—and formed column, *Cyane* half a cable's length astern of *Levant. Constitution* took station to their starboard and about two hundred yards abeam of the British ships and an exchange of broadsides began. Stewart maintained the weather gauge, allowing him to choose the range.

Cyane carried twenty-two 32-pounder and ten 18-pounder carronades and two 12-pounder long guns, and *Levant* eighteen 32-pounder and one 12-pounder carronades and two long 9s, for a combined throw-weight of 1,514 pounds. *Constitution's* battery consisted of twenty 32-pounder carronades, two 24-pounder "shifting gunades" and thirty long 24s, amounting to a throw-weight of 1,408 pounds. At the time of the engagement, *Cyane* and *Levant* had roughly 180 and 150 men aboard, respectively, while *Constitution*'s complement appears to have been 451. *Constitution* was both bigger and more sturdily built than either English ship and, though its battery was slightly smaller in throw-weight than the aggregate of *Cyane* and *Levant,* it heavily outgunned either ship individually. The British ships were armed chiefly with carronades, and the only long guns each of the British ships carried were two chase guns each.

Constitution's last War of 1812 commander, Charles Stewart, had a distinguished naval career of more than sixty years. *U.S. Naval Academy Museum*

A ball from the first British broadside slew two men near Stewart's station. After an intense exchange of broadsides, Stewart ordered firing stopped in order to let the smoke clear. He then saw that *Constitution* had drawn ahead, parallel with *Levant,* and that *Cyane* was seizing the opportunity to attempt to luff up and rake *Constitution*'s stern. In a rapid succession of skillfully executed movements, Stewart outmaneuvered his two opponents and insured victory over both. Firing a broadside into *Levant*, he ordered sails backed, bringing *Constitution* abreast of *Cyane,* into which *Constitution* unleashed another broadside. As *Levant* turned to starboard to gain a raking position across *Constitution*'s bow, Stewart wore ship to port, passing between his opponents, and raked *Levant*'s stern. While *Levant* disappeared into the darkness to make repairs and restore order among the gun crews, *Constitution* continued its turn to come under *Cyane*'s port quarter and stern. Outmaneuvered, outgunned, and—with the range down to fifty yards—unable to flee, Falcon struck his colors at 6:45 p.m.

By the time *Constitution* took control of *Cyane,* placed a prize crew on board, and transferred the captured officers, *Levant* completed a wide circle by which it returned to support *Cyane.* At 8:40 p.m. the two antagonists passed within fifty yards, starboard to starboard, and exchanged broadsides. The outclassed sloop then turned to flee, received a stern rake in the process, and after a chase into the night also surrendered shortly after 10 p.m.

In the engagement, *Constitution* suffered three killed and twelve wounded, three of them mortally. Stewart reported combined British losses as thirty-five killed and forty-two wounded. As would be expected when a heavy frigate fought much lighter vessels, the extent of damage to the rigging, spars, and hulls of the opposing sides was equally disproportionate.

Proceeding to Porto Praia in the Cape Verdes, Stewart turned the English prisoners over to the Portuguese and repaired all three ships in preparation for sailing home. Before he could depart, however, three British frigates hove in view out of a fog to investigate the harbor. Not trusting the British to honor the neutrality of the port, Stewart had his squadron under way within fifteen minutes. Having cleared the harbor, Stewart detached *Cyane* in the hope that the enemy squadron would pursue the easier, but faster, prey. The British frigates stayed on

Stewart's course, however, and closed the range enough to fire a few broadsides through the fog. As Stewart later wrote, "It became necessary to separate from the *Levant* or to risk being brought into action to cover her." The whole enemy squadron tacked to pursue *Levant*—perhaps owing to the fog's obscuring *Constitution*'s size and identity—and Stewart escaped. The sloop was not so fortunate and was recaptured.

When he put into Puerto Rico, Stewart learned that the Treaty of Ghent, putting an end to the war, had been ratified on 17 February. Although the ratification had taken place three days before *Constitution's* encounter with *Cyane* and *Levant,* the fight was well within the period after ratification that the negotiators, recognizing the difficulties of communication, had allowed for legitimate captures. After refitting and waiting for new orders, Stewart set course for home and arrived in New York on 16 May, where he received a hero's welcome.

With the War of 1812 concluded, in recognition of *Constitution's* remarkable victories during that conflict over *Guerriere, Java,* and *Cyane* and *Levant,* the *National Intelligencer* proposed formally recognizing the revered warship's status as a national icon:

> *"Old Iron Sides" at home. She has, literally become a Nation's Ship, and should be preserved. Not as a "sheer hulk, in ordinary" (for she is no ordinary vessel); but, in honorary pomp, as a glorious Monument of her own, and our other National Victories.*

For related documents, see pp. 106–14.

MJC

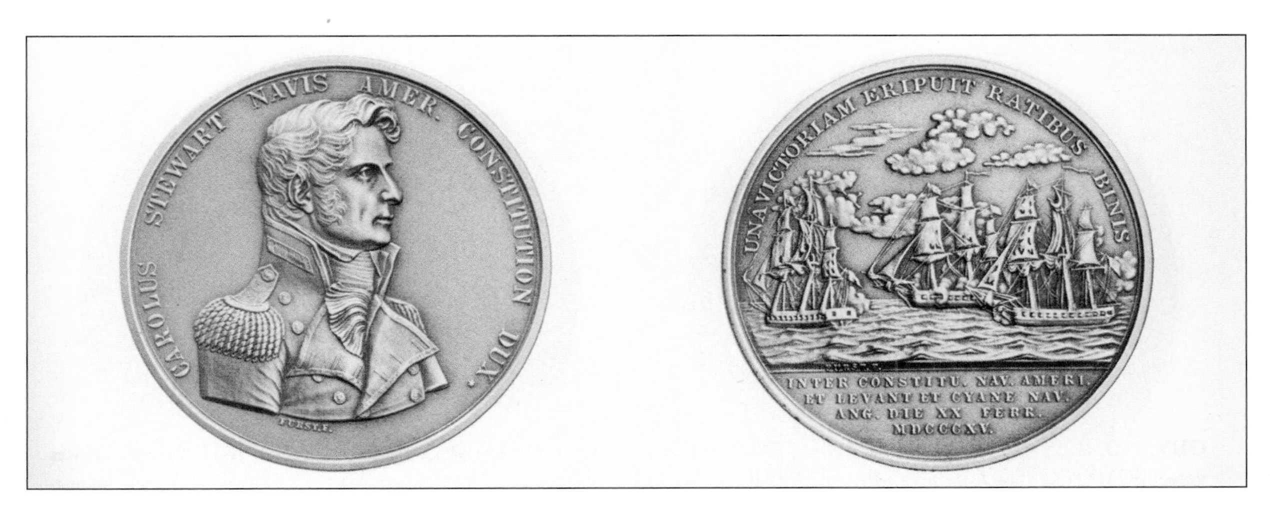

Congress awarded Charles Stewart a gold medal for his gallant conduct in the engagement with *Cyane* and *Levant*. The Latin inscription on the medal's upper reverse proclaims: "He snatched victory from two vessels with one." *Naval Historical Center photograph*

my station at quarters, or in time of battle, was as powder boy at gun No. 36, on the main gun-deck; my hammock number was six hundred and thirty nine; my ship's number, five hundred and seventy-four; and the number of my mess, twenty-six. Thus was the whole routine of my life on board this vessel laid out for me.

Part II: **Advanced Level**

Sister Ships: *United States* and *President*

Authorized in 1794, *Constitution, United States,* and *President* constituted a new class of 44-gun super-frigates, designed by Philadelphia shipwright and naval constructor Joshua Humphreys to meet the strategic needs of the new American republic. Unable to build a fleet large enough to compete with the powerful navies of Europe, the United States chose to build a few mighty warships with thick hulls capable of carrying heavy armament and withstanding the impact of the cannon shot of the weapons of conventional frigates. Built with the timbers of a ship of the line, but with the sleek lines of a frigate, these new ships of war proved themselves capable of defeating opponents of the standard frigate class and swift enough to escape more powerful ships and squadrons.

USS United States

Built at Philadelphia by Joshua Humphreys, *United States* was first of the 44s to enter the water, on 10 May 1797. John Barry, Captain, USN, commissioned the frigate 11 July. A year later, at the start of the Quasi-War, Barry sailed *United States* to the Caribbean as the flagship of a squadron under orders to protect American commerce and authorized to capture armed French vessels. In November 1799, *United States* carried to Europe commissioners appointed to settle the dispute with France. The frigate had resumed duty as flagship in the Caribbean in the spring of 1801 when the ratification of a treaty with France brought about its recall.

United States was laid up in ordinary at the Washington Navy Yard until 1809, when it was ordered into active service under Captain Stephen Decatur. Reportedly, while the frigate was refitting at Norfolk, Virginia, Captain John S. Carden, RN, of the new frigate HMS *Macedonian,* wagered Captain Decatur a beaver hat that his vessel would take *United States* if the two should ever meet in battle.

A few months into the War of 1812, the seemingly fated encounter settled the bet. In October 1812, *United States,* under command of Stephen Decatur, met Carden's *Macedonian,* sailing alone, five hundred miles south of the Azores. Decatur fought at a distance to take advantage of *United States*'s superiority in long guns (24-pounders against *Macedonian*'s 18s). The American frigate's broadsides damaged *Macedonian*'s hull and made a wreck of its rigging, rendering the Britisher's ability to maneuver severely impaired. *Macedonian*'s loss in men was forty-three killed or mortally wounded, and sixty wounded not mortally, whereas in contrast *United States* lost seven killed, with five more wounded. *United States* accompanied its prize into New York. Later, *United States* took refuge in New London, Connecticut, where the British blockaded it until the end of the war.

Sailing to the Mediterranean in 1815, *United States* arrived too late to share in the glory of forcing the dey of Algiers to make peace, but remained with the Mediterranean Squadron until 1819. In succeeding decades, between periods of repair and inactivity, the aging frigate served in the Pacific, Mediterranean, Home, and Africa Squadrons. Herman Melville, who would later write *Moby Dick,* based his novel *White-Jacket* on his experiences as an able seaman in *United States* during a voyage from Hawaii to New York. Although its original officers admired the frigate's speed under sail, its dull sailing eventually earned *United States* the nickname "Old Wagon."

The outbreak of the Civil War found the "Old Wagon" rotting at the piers of the Gosport Navy Yard. Confederates captured the yard and, renaming the ancient frigate CSS *United States* (frequently cited as CSS

In this woodcut, Sailing Master William Brady, USN, depicted the frigate *United States* under full sail. *Naval Historical Center photograph*

Confederate States), armed it with a deck battery for harbor defense. The Confederates sank the frigate in the Elizabeth River when they abandoned Norfolk, and the federal authorities raised it and towed it back to the yard, where it was broken up after the end of the hostilities in 1865.

The frigate *President*, shown here weathering a storm off Marseilles, surrendered to a British squadron in January 1815, while attempting to escape the enemy's blockade off New York. *Navy Art Collection, Naval Historical Center*

USS President

The construction of USS *President* began at New York, with Forman Cheeseman as constructor and under the superintendence of Captain Silas Talbot. Temporarily suspended in 1796 when Algiers ended its hostilities against the United States, construction resumed under the superintendence of Lieutenant Isaac Chauncey in 1798 at the beginning of the Quasi-War with France. *President* was launched on 10 April 1800 and sailed on its first cruise on 5 August, Captain Thomas Truxtun in command, serving as flagship of the squadron patrolling the West Indies until news of a peace treaty with France arrived at the end of the year.

Commodore Richard Dale employed *President* as his flagship during the first years of the War with Tripoli in 1801 and 1802, and *President* returned to the Mediterranean to participate in the culminating campaign against Tripoli in 1804 and 1805.

From 1810 to 1812 *President* was flagship of Commodore John Rodgers's squadron, assigned the duty of patrolling the northeastern coast of the United States. On the night of 16 May 1811, *President* encountered an unknown warship that Rodgers took for a frigate. Before the two ships identified themselves to each other, one of the two ships, it is not clear which, fired a single cannon. A gun division in *President* fired in return and a general engagement followed. After fifteen minutes the stranger's gunfire was silenced. In the aftermath, Rodgers learned that *President*'s opponent was HMS *Little Belt,* a corvette of only twenty guns. Some Americans found the encounter embarrassing, from the point of view of the discrepancy in force between the two vessels; others viewed it as retribution for the 1807 *Chesapeake-Leopard* affair, in which a British ship fired broadsides into a U.S. Navy frigate before removing from it several Royal Navy deserters.

With the outbreak of the War of 1812, *President* continued as Rodgers's flagship. During the squadron's first cruise it encountered and chased H.M. frigate *Belvidera,* which escaped after a gun on board *President* exploded, injuring Rodgers. During a cruise in September 1813, *President* captured H.M. schooner tender *Highflyer,* of only five guns.

During the winter of 1813 and 1814 *President* cruised in the West Indies, and returning to New York in February 1814 was blockaded there until 14 January 1815, when Stephen Decatur attempted to take it out to sea. *President*'s rudder and keel were injured on a shoal in the attempt, and, unable to escape, *President* engaged H.M. frigate *Endymion* until *Endymion*'s consorts, *Majestic, Pomone,* and *Tenedos* came up and obliged Decatur to submit. The British took *President* into the Royal Navy, retaining it until 1817, when they broke the frigate up at Portsmouth.

MJC

The Quasi-War and U.S. frigate *Constitution*

In defense of American commerce, on 28 May 1798 Congress authorized the Navy to capture armed French vessels off the coast of the United States and on 9 July extended the authorization to anywhere on the high seas. The first seizure of a French vessel during the Quasi-War was in Delaware Bay, but most of the action of the undeclared war took place in the West Indies. There, as directed by Secretary of the Navy Benjamin Stoddert, U.S. naval squadrons patrolled for French armed ships and conducted convoys of American merchantmen. The conflict lasted until December 1800, when news reached Washington, D.C., that France had agreed to a peace treaty. The Senate's ratification on 3 February 1801 of the new peace accord, known as the Convention of Mortefontaine, brought hostilities between the two countries to a formal close.

In the course of the Quasi-War, American ships made prizes of approximately eighty-five French vessels. Two of the most significant naval actions involved U.S. frigate *Constellation,* under command of Captain Thomas Truxtun. In February 1799 *Constellation* captured French frigate *Insurgente* and a year later fought to a draw the more powerful *Vengeance*.

Constitution's shakedown cruise began on 22 July 1798, as the frigate, under command of Captain Samuel Nicholson, departed Boston Harbor under orders to protect the principal ports of New England and New York. In August, the warship turned south, with orders to cruise between the Virginia Capes and the coast of Georgia. On 8 September off North Carolina, *Constitution* made its first capture, the 24-gun armed sloop *Niger,* whose captain asserted that he and most of his crew were French Royalists operating against the French and Spanish under a British letter of marque. An American admiralty court later determined that *Niger* was British property and ordered the sloop returned to its owners, with monetary damages.

Constitution sailed again from Boston on 29 December 1798 and set course for the West Indies. On 17 January 1799 *Constitution* captured *Spencer*, an English merchantman recently taken by the French frigate *Insurgente* while enroute to Barbados. Made cautious by his experience with *Niger,* Nicholson decided that his orders authorized him to make prize of armed French ships but not of French prizes and let *Spencer* go.

Nicholson's errors in judgment destroyed Secretary Stoddert's confidence in him, and when *Constitution* returned to Boston in May, Captain Silas Talbot relieved Nicholson in command. In July Talbot sailed under orders putting him in command of a squadron stationed off Haiti that was to lend support to the Haitian independence movement led by a former slave, François Toussaint L'Ouverture, and protect American commerce from attack by freebooters, particularly French privateers.

On 9 May 1800, U.S. schooner *Experiment* and a detachment of seamen and Marines in four of *Constitution*'s boats took from a minor Haitian bay the 6-gun French privateer *Esther* with the captured Massachusetts

Cutting out expeditions, such as that of USS *Constitution*'s Marines and Sailors boarding and seizing the French privateer *Sandwich* at Puerto Plata on 11 May 1800 depicted here, were among the most dangerous and difficult operations to execute successfully in the age of sail. *Boston Athenæum*

As the first secretary of the newly established Navy Department, Maryland merchant Benjamin Stoddert skillfully directed American naval operations during the Quasi-War with France. *Navy Art Collection, Naval Historical Center*

merchant brig *Nymph* in a brisk half-hour action.

At midday two days later the American merchant sloop *Sally* entered Puerto Plata, a small port on the northern coast of Hispaniola, and ran lightly alongside the French privateer corvette *Sandwich*. Concealed on board *Sally* was a cutting out party of eighty Marines and seamen, commanded by Isaac Hull, one of *Constitution*'s lieutenants. The boarding party quickly overwhelmed *Sandwich* and took it prize. Defended from sporadic French counterattacks by U.S. Marines who had spiked the guns of the battery above the privateer, Hull and his seamen readied *Sandwich* for sea. *Sally* and its prize joined *Constitution* at sea the next day. The United States subsequently acquiesced to Spain's protests against this action as a violation of its neutrality by returning *Sandwich* to its French owners.

On 25 August 1800 *Constitution* returned to Boston and departed on 17 December for one more cruise in the West Indies while the diplomatic process was bringing about a peace settlement.

MJC

The Barbary Wars

During the colonial period, American merchants developed a flourishing trade with Mediterranean countries, taking advantage of the protection from the corsairs of the Barbary Powers afforded by the British navy. For centuries, the Islamic governments of Morocco, Algiers, Tunis, and Tripoli had demanded tribute from Christian nations in return for safe passage through the sea. Soon after the War of Independence, American merchants were dispatching some one hundred ships, manned by twelve hundred seamen, to the Mediterranean each year, but American ships no longer enjoyed the protection of the British flag and the United States had no navy of its own to protect its commerce. By 1784 Morocco and Algiers had both captured American merchantmen and held their crews captive. A treaty was soon worked out with Morocco calling for small payments of tribute, but Algiers was looking for substantial remuneration. Lacking the funds either to use force or to pay tribute, Congress did nothing and the United States remained at war with Algiers.

Ratified in 1789, the Constitution gave the federal government power to tax and to create a navy. In 1794 the United States began building warships to deal with the Barbary problem, and the following year American negotiators agreed to pay Algiers an annual tribute. Algiers then helped the United States arrange treaties with Tripoli and Tunis.

Soon the pasha of Tripoli grew dissatisfied with the level of tribute and threatened war. In 1801 President Thomas Jefferson sent Commodore Richard Dale with a squadron to the Mediterranean. If any of the Barbary Powers had declared war against the United States, Dale was to blockade their ports and destroy their ships. When Dale arrived in the Mediterranean, he learned that the Pasha of Tripoli had declared war.

Dale attempted to blockade the harbor of Tripoli with his three frigates and one schooner but found that he needed more ships of shallow draft. Richard Valentine Morris replaced Dale in May of 1802, with one additional frigate under his command. Instead of focusing on the blockade, which had the potential of bringing the pasha to negotiate, Morris focused on convoying trade, which might protect merchantmen but would not end the war. Unhappy with the progress of the war, the Navy Department recalled Morris in the summer of 1803 and dismissed him from the service.

Morris's replacement, Edward Preble, arrived in the Mediterranean in his flagship, U.S. frigate *Constitution,* determined to bring about a peace without tribute. He intended to do so not by blockade alone but also by bombarding Tripoli's capital city into submission.

Preble's hopes of ending the war in one season, however, vanished when the U.S. frigate *Philadelphia* went aground on a reef east of Tripoli Harbor. Surrounded by Tripolitan gunboats, the frigate surrendered. When the rising tide lifted *Philadelphia* free, the Tripolitans towed it into harbor. The pasha now had a fine frigate and 307 U.S. naval officers and men in custody for whom he could demand ransom.

Convinced that it would be impossible to retake the *Philadelphia* and sail it out of the harbor under the guns of the fortress, Preble decided to destroy it instead. In a captured ketch renamed *Intrepid* and disguised as a peaceful trader, volunteers from *Constitution* and U.S. schooner *Enterprise* under command of Stephen Decatur entered the harbor, boarded the frigate, and set it ablaze.

On 3 August 1804, Preble began a vigorous campaign of bombardment of the city of Tripoli. American gunboats engaged the enemy fleet while *Constitution* kept up a constant fire on the enemy shipping and shore batteries. The U.S. squadron repeated the bombardment on 7, 23, and 27 August, and 2 September. Then on the night of 3 September, a volunteer crew of twelve in *Intrepid,* converted into a fire ship, attempted to enter the harbor to blow up the shipping. *Intrepid* exploded prematurely, killing all on board.

Samuel Barron relieved Preble on 9 September. Preble returned to America to accolades and a gold medal voted by Congress. Arriving with a reinforcement of five frigates, Barron tightened the blockade but did not resume bombardment. Command of *Constitution* passed to John Rodgers.

William Eaton, U.S. Navy Agent to the Barbary Powers, planned to bring the war to a satisfactory conclusion by replacing the reigning pasha, Yusuf Karamanli, with his brother, Ahmed. With Eaton at its head as general, Ahmed's army, consisting of men from eleven nationalities, three hundred mounted Arabs, seventy Christians, a midshipman from USS *Argus,* and eight U.S. Marines, set out from Alexandria, Egypt, on 6 March 1805. The little army reached the outskirts of Derna on 26 April and the next day, supported by brig *Argus,* schooner *Nautilus,* and sloop *Hornet,* captured the city.

In the meantime, Tobias Lear, U.S. consul at Algiers, judging that a peace won from Yusuf would be more secure than one guaranteed by Ahmed, a weakling who would not be able to retain power, was working a rival plan to bring about peace. On 3 June, the pasha, shaken by the internal support emerging for his brother and impressed by the power of the augmented U.S. squadron and the effectiveness of the blockade, signed a preliminary peace negotiated by Lear, and the next day released the *Philadelphia*'s officers and crew. On 10 June the final peace treaty was signed, ending tribute payments and providing that America pay a ransom for the release of the *Philadelphia*'s crew.

Constitution remained on duty in the Mediterranean until the autumn of 1807. When it returned to Boston after its four-year deployment, it carried home

In this first (3 August) in a series of five bombardments in August and September 1804, an American squadron, covered by Edward Preble's flagship *Constitution*, sought to force the Tripolitans to the peace table by destroying their capital's shore and naval defenses. *U.S. Naval Academy Museum*

a marble monument, a tribute to the naval heroes who gave their lives in the Tripolitan war.

The Barbary powers became less troublesome, but fully peaceful relations were not established until after the War of 1812. During that war Algiers sided with Great Britain and near its end renewed its attacks on American merchantmen. On 2 March 1815, having ratified the Treaty of Ghent with Great Britain, Congress declared war on Algiers. Stephen Decatur returned to the Mediterranean with a squadron of three frigates and seven smaller vessels, and, before the harbor of Algiers, dictated the terms of peace from the mouths of his ships' cannon. The dey of Algiers released his American prisoners, paid an indemnity for the depredations of his corsairs, and gave up all claims to future tribute. The war's surprisingly rapid end resulted in cancellation of *Constitution*'s orders to the Mediterranean to join Decatur's squadron.

MJC

Edward Preble and Preble's Boys

Edward Preble

Edward Preble (1761–1807) of Maine served as an officer in the Massachusetts State Navy during the Revolution, and, captured by the British, was held for a time in the infamous prison hulk *New Jersey*. After the Revolution he served fifteen years as a merchant sea captain until 1798, when he was appointed lieutenant in the U.S. Navy. He commanded the 14-gun brig *Pickering* in the West Indies during the Quasi-War. Commissioned captain in 1799, he took the frigate *Essex* into the Pacific to protect America's East Indies trade. With *Constitution* as his flagship, in 1803, he sailed for the Barbary Coast and by October had promoted a treaty with Morocco and established a blockade off Tripoli. Relieved in September 1804, Commodore Preble returned to the United States and took up shipbuilding at Portland, Maine.

Stephen Decatur leads his crew in a desperate fight for control of an enemy gunboat during the American attack on Tripoli, 3 August 1804. "I find hand to hand is not childs play," Decatur would later remark, "'tis kill or be killed." *Navy Art Collection, Naval Historical Center*

Daniel T. Patterson

Edward Preble

James Biddle

Thomas Macdonough

In the early Navy, captains such as Edward Preble mentored aspiring younger officers such as Biddle, Macdonough, and Patterson, who subsequently distinguished themselves during the War of 1812. *Preble, Biddle, and Macdonough: U.S. Naval Academy Museum; Patterson: Chrysler Museum of Art*

Preble's Boys

Many of the officers who led the Navy during the War of 1812 had been lieutenants during the War with Tripoli. During the earlier war they gained experience in daring exploits and close action under the leadership of Commodore Edward Preble. The youthfulness of these young lieutenants, it is said, led Preble to fume that he had been given "nothing but a pack of boys"—a more reliable source has it, "nothing but a parcel of children." Preble's lieutenants are supposed to have nurtured their aggressive spirit and formed their tactical ideas under the powerful influence of that fighting Sailor, taut disciplinarian, and irritable personality. From this supposition comes the notion that the Navy's most successful leaders in the War of 1812 were "Preble's Boys," a phrase popularized by Fletcher Pratt's 1950 book of the same name, *Preble's Boys: Commodore Preble and the Birth of American Sea Power.*

In Pratt's estimation, the term *Preble's Boys* embraces William Bainbridge, James Biddle, Johnston Blakeley, William Burrows, Stephen Cassin, Isaac Chauncey, Stephen Decatur, Isaac Hull, Jacob Jones, James Lawrence, Thomas Macdonough, Daniel Todd Patterson, David Porter, Charles Stewart, and Lewis Warrington. "With the single exception of the Battle of Lake Erie," writes Pratt, "every victory in the War of 1812 was won by one of Preble's boys. With three exceptions, every one of Preble's boys who had a command in 1812 brought home at least one British battle-flag."

Christopher McKee's scholarly statistical analysis of the ages, experience, and ranks of the pool of naval officers in the War of 1812 demonstrates that Platt's thesis is more art than science. McKee concludes that the men who served under Preble in the Tripolitan War won the Navy's battles in the War of 1812 not because they had learned from Preble's example, but because, owing to their seniority and other reasons, they were the officers who held fighting commands in the later conflict. "It was absolutely impossible for any captain who was not a Preble veteran to win a victory at sea in the War of 1812, for there was not one afloat!" Besides, all the naval officers of the War of 1812 who had served under Preble had also served under other officer role models, Thomas Truxtun, John Rodgers, William Bainbridge, and others. McKee exhorts us to avoid the "simplistic approach that has dominated and distorted the study of the U.S. Navy's history. One speaks here of the focus on individual combat commanders, often accompanied by the implied or explicit claim that the officer in question is *the* premier shaper of the American naval tradition." [1]

If not the chief cause, Preble's example of leadership certainly contributed to the development of the naval officer corps into a highly competent body. In McKee's words, Preble's moral virtues—readiness to spend himself for his country, holding himself and subordinates to the highest standards of professional conduct and ethics, willingness to delegate responsible assignments, and vigorous action in the face of reverses—"made him the preeminent mythic hero and model to the officer corps of the War of 1812."

1. Christopher McKee, "Edward Preble and the 'Boys': The Officer Corps of 1812 Revisited," in *Command Under Sail: Makers of the American Naval Tradition, 1775–1850*, edited by James C. Bradford (Annapolis, MD: Naval Institute Press, 1985), 71–96.

MJC

War of 1812: The Navy Department

When the War of 1812 began, the Department of the Navy was fourteen years old, having been established in 1798. Its offices were co-located with those of the Departments of State and of War in a building on the north side of Pennsylvania Avenue, between 21st and 22nd Streets, N.W., in Washington, D.C., some two hundred yards west of the President's House (the White House). After the British burned the building in August 1814, the department moved into leased quarters on the corner of 20th and I Streets, N.W.

Presiding over the department in 1812 was Paul Hamilton, a former governor of South Carolina, who had held the post of secretary of the navy since 1809. It has been said that his inability to handle the responsibilities of running the department under the heavy demands of wartime drove him to alcoholism, which led to the president's asking him for his resignation on 29 December 1812. It is just as likely, however, that it was simply his incompetence as an administrator—he did not keep track of expenditures, appointed more midshipmen than authorized, let contracts to friends—that caused Congress to lose confidence in him and demand Madison replace him.

In William Jones, who held the office of secretary of the navy from 19 January 1813 until 1 December 1814, the country found an efficient and effective administrator. A Philadelphia merchant and former congressman, Jones had sailed as an officer in privateers during the American Revolution and also as a merchant sea captain. Jones imposed order and method to the department. He was blunt and correct in his relations with Navy officers, but not warm. He resigned the post because of the need to address personal financial difficulties.

Jones advocated attention to the requirements of operations on the northern lakes, to the chagrin of coastal dwellers, who objected to the drain of resources from their own defense. He also resisted the demands of every port and harbor for gunboat protection, refusing to scatter the available forces. On the oceans, Jones favored commerce raiding over operations against the Royal Navy, and the smaller classes of warships that could more easily evade the British blockade.

Benjamin Williams Crowninshield, a merchant and privateer owner of Salem, Massachusetts, who had formerly commanded East Indiamen, accepted the position

Paul Hamilton, secretary of the navy at the outset of the War of 1812, resigned six months after hostilities commenced, unable to manage a complex wartime bureaucracy. *Navy Art Collection, Naval Historical Center*

As Paul Hamilton's successor, William Jones established an effective war machine with limited resources during his tenure as secretary of the navy. *Navy Art Collection, Naval Historical Center*

Benjamin Crowninshield replaced William Jones as Navy secretary during the last month of the war and oversaw the Navy's transition to peacetime service. *Navy Art Collection, Naval Historical Center*

of secretary of the navy, on 16 January 1815, just as the war was concluding.

The chief clerk of the department had responsibility for the navy's records, managing the flow of business and correspondence, drafting replies to letters, communication between the secretary and officers, controlling visitors' access to the secretary, and drafting reports and annual estimates for Congress. In 1812, the chief clerk had three assistants; by the close of the war that number had increased to six full-time and one part-time assistant clerks.

While gunboats failed to realize their promise as guardians of America's coastal waters during the war with Great Britain, they did perform important convoy, transport, and reconnaissance services. *The Pictorial Field-Book of the War of 1812, Vol. 1*

The office of the accountant of the navy consisted of the chief accountant and from nine to twelve assistant accountants. It conducted the financial business of the navy, the most important aspect of which was the review and settlement of accounts.

The department entrusted responsibility for administrating the navy yards and stations to navy captains. Civilian navy agents appointed in the principal ports made purchases for the navy, for a commission. Naval constructors built the navy's warships, under the superintendence of naval officers.

In 1815, after the end of the war, Congress created the Board of Navy Commissioners, consisting of three active naval officers, to provide professional advice to the secretary and relieve him of some of his administrative burden, which the war had greatly increased.

MJC

War of 1812: Overview of the U.S. Fleet

On the eve of the War of 1812 the fleet in service consisted of fifteen warships in active commission, the largest of which were frigates of forty-four guns, and sixty-two gunboats. Laid up in ordinary were a handful of frigates and eighty-six gunboats, with another seven gunboats undergoing repairs. The Navy operated yards at Portsmouth, New Hampshire, Charlestown, Massachusetts, Brooklyn, New York, Philadelphia, Pennsylvania, Washington, D.C., and Gosport, Virginia. In addition there were small naval stations at such locations as: Sackets Harbor, New York, on Lake Ontario; Newport, Rhode Island; Wilmington, North Carolina; Charleston, South Carolina; St. Marys, Georgia, on the border with Spanish Florida; and New Orleans, Louisiana. The service's total manpower stood at about one thousand officers and men, with only thirteen captains, the highest rank in the Navy before the Civil War. On the outbreak of war, the Navy began procuring additional ships and recruiting Sailors and Marines to man them.

Despite the Navy's initial successes on the high seas in the first months of the war, U.S. policymakers believed that ships of the line would be essential if the American navy was ever to compete meaningfully against British sea power. On 2 January 1813 President Madison signed an act providing for the building and fitting out of four 74-gun ships and six 44-gun frigates. The new Navy secretary, William Jones, however, saw the money and material devoted to the building of ships of the line as misdirected. He doubted large vessels could be completed in time to help the war effort. Such ships would be restricted to the larger ports with sufficient depth of water, and these ports were vulnerable to the British blockade. Smaller ships could be completed much more quickly and could make use of the smaller harbors, too numerous for the British to blockade completely. Shortly after Jones took office, Congress authorized the building of six sloops of war for ocean service. On 16 November 1814, the day after Congress authorized the purchase or construction of up to twenty vessels armed with between eight and twenty guns, Jones ordered the acquisition of schooners to form "flying squadrons" to harass British commerce. At

During the War of 1812 the Navy Department contracted with inventor Robert Fulton to build the steam-powered floating battery *Demologos* or *Fulton I* to aid in New York Harbor's defense. This scene depicts the steam frigate's launching at New York on 29 October 1814. *Naval Historical Center photograph*

war's end, the Navy had a steam-powered vessel for the defense of New York Harbor.

After two years of war, the Navy still had no ships of the line in service, but had otherwise grown substantially, with seventeen frigates and ships, more than four dozen mid-size sailing armed vessels (schooners, brigs, sloops), nearly four dozen barges, a dozen galleys, and a scattering of other vessel types, besides the one hundred and one gunboats in service. (See "Distribution of U.S. Naval Forces, 1 June 1814" below.) In October 1814, the U.S. Navy had 10,600 men in its ranks. The majority of these, 6,500, were on the coast employed in harbor defense and in flotillas for the protection of the Chesapeake Bay. Some 3,250 were on the northern lakes. A mere 450 were at sea, in only three ships. Another 405 were in British prisons.

The Sailors stationed on the coast manned flotillas of gunboats, barges, and other small craft in the nation's principal harbors from Maine to Louisiana. Defense was the principal duty of the gunboats, twenty-five of which the Navy lost during the war.

The transfer of men and ordnance to the lakes hampered coastal defense. In contrast to the ocean, on the lakes the United States had to match the British ship for ship and man for man. As the British transferred more and more sailors and soldiers from the European theater to the lakes, the American navy began to run short of seamen for new ships and of guns to arm them. In October 1814, Secretary Jones estimated the need on the lakes for another 3,500 men, more than the number already there. Of eight ships ready for sea in 1814, four had been stripped of their crews in order to man the lake fleets. And it would be necessary to remove the guns of the new ships of the line *Independence* and *Washington* to arm the ships planned for Lake Ontario.

The war made manifest the importance of the oceans to America. Even Jeffersonians now recognized the need to be strong at sea. At the end of the war there was no talk about putting the Navy back into ordinary, as had been done following the wars with France and Tripoli. Rather, the discussion was about how to augment the Navy. In 1816 Congress provided for its gradual increase through the expenditure of one million dollars a year for six years to build nine 74-gun ships of the line, twelve 44-gun frigates and three steam-propelled batteries for harbor defense.

MJC

Distribution of U.S. Naval Forces, 1 June 1814

Atlantic Frontier

Vessel Type	On the Coast	At Sea	Ready for Sea	Wanting Men	In Ordinary	Building
Ships of the line						3
Frigates and Ships	1		3[1]		3	3
Sloops and Corvettes	1	3	2	1	2	
Brigs		2				
Schooners	9					1
Gunboats	101				10	
Barges	34			10	3	
Bomb vessels	2					
Block and Guard Ships	5					1
Cutters	1					
Feluccas	1					
Galleys	2					
Hospital Ships	1					
Scows					3	
Steam Battery						1
Totals	**158**	**5**	**5**	**11**	**21**	**9**

1. Includes one frigate reported as being nearly ready for sea.

Northern Lakes

Vessel Type	Lake Champlain	Lake Ontario	Lake Erie
Ships	1	4[1]	2[2]
Sloops	2		
Brigs		4	4
Schooners	1	11	8
Galleys	10		
Totals	**14**	**19**	**14**

1. Includes one ship building.
2. Both ships in ordinary undergoing repair.

The figures in these tables are based on Secretary Jones's report of 6 June 1814 to President Madison published in *The Naval War of 1812: A Documentary History*, 3:780–87.

War of 1812: U.S. Naval Operations

On the Oceans

1812

At the outset of the War of 1812, Secretary of the Navy Paul Hamilton ordered Commodores John Rodgers and Stephen Decatur to have their two small squadrons cruise off their respective stations, New York and Norfolk, staying close to the coast to protect merchant vessels returning to their home ports and joining forces only when they expected to meet a superior force. The day before Hamilton wrote these orders, Rodgers, with his four-ship squadron, sailed on a cruise that extended far across the Atlantic. Although Rodgers achieved little success in capturing prizes, his cruise caused the British admiral in command of the North American Station to dispatch a similarly sized squadron to protect commerce, thus occupying ships of the Royal Navy that otherwise would have blockaded American ports.

The U.S. fleet lost its first ship of the war when the brig *Nautilus* fell in with the enemy North American squadron on 15 July. Cruising independently the frigate *Essex* captured H.M. sloop of war *Alert,* and the sloop *Wasp* defeated H.M. brig-sloop *Frolic* before being overtaken and captured by H.M. ship of the line *Poictiers.*

In the fall of 1812, after Rodgers's return, the Navy Department formed three small squadrons, under Rodgers, Decatur, and William Bainbridge, each of which had a measure of success. Rodgers's squadron captured a British Post Office packet boat carrying a cargo of gold and silver coin; Decatur's frigate *United States* captured the new British frigate *Macedonian,* and Bainbridge's *Constitution* won its second victory of the year over a British frigate.

1813

At the turn of the new year, a new secretary, William Jones, took over the Navy Department. Jones believed that the most effective possible use of the navy was raiding enemy commerce. He favored expanding the Navy with small raiders rather than with large ships of the line.

During 1813, the British tightened their blockade of the American coast. Frigates *President* and *Congress* slipped through the blockade for separate cruises. The frigate *Essex* sailed around Cape Horn into the Pacific where the British whaling fleet operated unprotected. At year's end, *Constitution* left port for its third wartime cruise. But other frigates were unable to leave port, and *Chesapeake,* attempting to escape from Boston, fell victim to the superior gunnery of HMS *Shannon.*

Smaller American warships met various degrees of success on the ocean. U.S. sloop of war *Hornet* captured H.M. brig-sloop *Peacock* in an engagement off British Guiana; U.S. brig *Argus* took nineteen prizes off England before being captured by H.M. brig-sloop *Pelican;* and U.S. brig *Enterprise* took H.M. gun-brig *Boxer* off the coast of Maine.

A British squadron was able to dominate the Chesapeake Bay and launch a number of raids on bay towns in order to relieve pressure on the Canadian border. In June, at the Battle of Craney Island, U.S. frigate *Constellation* and the Virginia militia repelled a British assault aimed at Norfolk.

1814

By the beginning of 1814, America's opportunity had been lost. After Napoleon's capitulation in April, more Royal Navy ships were available to close the blockade of

Having long guns and carronades that fired shot heavier than those mounted in *Macedonian* gave *United States* a decisive edge in its engagement with the British frigate in their 25 October 1812 engagement. *Navy Art Collection, Naval Historical Center*

A disaffected crew and poor gunnery contributed to H.M. brig-sloop *Epervier*'s loss to U.S. sloop of war *Peacock* on 29 April 1814. *Navy Art Collection, Naval Historical Center*

landed an army of four thousand men at Benedict, Maryland, and marched toward Washington. On 24 August they broke an ill-organized defense at Bladensburg, where, among the defenders, four hundred Sailors and Marines under Captain Joshua Barney manned a battery of 18-pounder guns. The British reached Washington at 8 p.m. They set fire to the Capitol, the President's House (White House), the Treasury, the War Office, and other public property. They left the city the next night and were back at Benedict on the 29th. A large contingent of U.S. Navy seamen assisted in manning shore cannon during the subsequent successful defense of Baltimore, 12–14 September.

the U.S. coast, and army reinforcements changed the role of the British forces in America from defensive to offensive. In the summer of 1814 the Royal Navy controlled the waters around the United States, throttling the small American navy and making possible a series of attacks on the American coast. U.S. brigs *Rattlesnake* and *Siren* both fell victims to enemy frigates they failed to outsail, and a brace of Royal Navy warships ended the *Essex*'s remarkable cruise wreaking havoc among the British whaling fleet in the Pacific.

Early in 1814 America launched three new sloops of war: *Frolic, Peacock,* and *Wasp*. *Frolic* was captured off Cuba. *Peacock* defeated H.M. brig-sloop *Epervier* off Cape Canaveral. *Wasp* raided commerce off the British coast and defeated H.M. brig-sloops *Reindeer* and *Avon* before disappearing in the Atlantic. Despite the exploits of *Peacock* and *Wasp,* vindicating Secretary Jones's faith in smaller warships, the U.S. Navy in 1814 made little impact on the oceans. Most of its vessels remained bottled up in port. In December, *Constitution* broke out once more for another cruise.

In the Chesapeake Bay in 1814, the British increased their raids on the coast towns. The United States government was unable to mount an effective defense against these depredations, as the British had complete naval control of the tidewater. On 19 August, they

During the summer the British began raids on the New England coast. In late summer the British took control of Maine from the New Brunswick border to the Penobscot River. In the resistance to this invasion, the American corvette *Adams,* of 24 guns, was burned to prevent capture.

1815

Before the news of the Treaty of Ghent became widely known, several naval events of note took place on the oceans. USS *Constitution* captured HMS *Cyane* and HMS *Levant* in a single action. The British blockading squadron captured U.S. frigate *President,* attempting to escape from New York. The sloops of war *Hornet* and *Peacock* escaped from New York, the former going on to capture H.M. brig-sloop *Penguin* and the latter the British East India Company brig *Nautilus*.

On the Lakes

1812

The year 1812 saw the disappointment of American expectations of a quick and easy conquest of Canada. The failure of all parts of a planned three-pronged attack, at Detroit, Niagara, and Montreal, resulted in large part from the neglect of naval concerns. Any suc-

cessful invasion of Canada would require naval control of Lakes Erie, Ontario, and Champlain.

At the beginning of the war, neither belligerent possessed much naval strength on the northern lakes. The Canadian Provincial Marine operated a handful of warships on the Great Lakes, and the U.S. Navy had a 16-gun brig on Lake Ontario and two dilapidated gunboats on Lake Champlain. Competition for mastery of the lakes during the war consisted largely of a race to build, outfit, and man more and larger ships.

By the end of 1812, Commodore Isaac Chauncey had won control of Lake Ontario with a small fleet, while the British continued in command of Lake Erie. Lieutenant Thomas Macdonough had begun to build a U.S. fleet on Lake Champlain.

Acting Midshipman Peter W. Spicer, who served on board U.S. schooner *Sylph*, sketched the 11 September 1813 Lake Ontario fleet action between the naval forces of Commodore Isaac Chauncey, USN, and Commodore Sir James Yeo, RN. *Navy Art Collection, Naval Historical Center*

1813

The Royal Navy assumed control of British naval forces on the northern lakes in the spring of 1813. Commodore Sir James Lucas Yeo's task was simpler than that of the American commander's. Yeo, with his headquarters at Kingston, Ontario, on Lake Ontario, had merely to defend against American invasion of Canada. Chauncey, stationed at Sackets Harbor, on the eastern end of Lake Ontario, had to win and maintain control of the lakes to support an invasion. In the spring, Chauncey's Lake Ontario fleet cooperated with American troops in a raid on York (present-day Toronto) and Fort George. While the American fleet was away, Yeo attacked Sackets Harbor. The assault was beaten off, but only after considerable damage was done to the American naval base. Through the rest of the season, the two fleets sailed around the lake engaging in mutual evasion. The U.S. squadron, with its majority of long guns, wanted to fight at long range. The British, with its superiority in carronades, wanted to fight close. The stalemate on Lake Ontario favored the British, since it stymied any American invasion.

On Lake Erie, Master Commandant Oliver Hazard Perry and his British counterpart, Captain Robert Barclay, each pressed forward with building warships. Perry sailed his fleet into the lake in August, obliging Barclay, with his more lightly armed ships, to sail to meet him in order to protect British supply lines. Perry announced the results of the Battle of Lake Erie, 10 September 1813, to the American army commander William Henry Harrison with the now famous line, "We have met the enemy and they are ours." On learning that the entire Royal Navy squadron had been captured, the British army commander ordered a retreat, during which the British army and its Indian allies were severely handled by the American forces under Harrison at the Battle of the Thames.

Not until November, with winter setting on, did the United States feel confident enough of its control of Lake Ontario to risk an invasion up the St. Lawrence River. The invasion was badly planned and executed, ending in a withdrawal.

1814

In 1814, the Americans exploited their control of Lake Erie by extending operations into Lake Huron. A joint campaign to retake Fort Michilimacinac failed, however, and after the withdrawal of the main American force, two U.S. schooners left behind to patrol Lake Huron fell to surprise attacks.

During the summer of 1814, with Napoleon exiled to the island of Elba, the British sent to the northern lakes reinforcements of men and ships in frame (for later assembly), and they continued their shipbuilding race with the Americans with the aim of taking the offensive. A British army of ten thousand crossed into the United States, along the western verge of Lake Champlain, on 1 September. The American force, fewer than four thousand men, made its defense at Plattsburgh, New York. Before attacking on land, the British believed it necessary to engage the American squadron positioned in Plattsburgh Bay. On 11 September, the British squadron attacked the American squadron under Macdonough anchored in a position of its choosing. Before the morning was out, the British ships had all struck their colors and their gunboats had fled. The British army retreated to the sound of the Americans cheering the naval victory.

In the Gulf of Mexico

1812–1813

Despite the crucial importance of the Mississippi River for exporting American produce, the U.S. naval station at New Orleans, Louisiana, was largely a backwater of the war during the conflict's first two years. Indians, pirates, and the Spanish concerned the station more than the British. In April 1813, naval forces escorted troop transports and blockaded Mobile Bay during the army's successful campaign to capture Mobile, in Spanish West Florida.

1814–1815

At the urging of the governor, in September 1814 an American naval force under Master Commandant Daniel T. Patterson attacked and broke up the base used by pirates and smugglers in Barataria Bay, on the southern coast of Louisiana.

From the beginning of the war, the British had recognized the importance of New Orleans, but not until late in 1814 were they ready to launch an operation there. The British chose to land troops at Bayou Bienvenu at the west end of Lake Borgne, fifteen miles from New Orleans. Because of the shallowness of the bayou, they had to anchor their fleet sixty miles away from the landing place and transport the troops in boats of shallow draft. Before debarking any troops, however, they had to defeat the American gunboat flotilla. On 12 December, the launches, barges, and pinnaces of the fleet rowed into Lake Borgne in search of the American gunboats. The battle, fought on the 14th, ended in the capture of five American gunboats and a sloop, and the burning of a schooner to prevent its capture. All these delays gave the American land forces more time to prepare a defense.

On the morning of 23 December, the British landed and advanced to within seven miles of the city of New Orleans, on a road that paralleled the Mississippi River. On the night of the 23d, the American troops under Andrew Jackson, supported by the U.S. Navy schooner *Carolina* and the ship *Louisiana* in the river, attacked the enemy force. The Americans then retreated two miles and set up a defensive line behind a shallow canal. The British destroyed *Carolina* with heated shot fired from a shore battery and forced *Louisiana* to retire. Sailors and Marines under Patterson fought in Jackson's lines and manned a battery on the western side of the Mississippi River that flanked those attacking the main American lines on the eastern side. On 8 January, a frontal assault against the American defensive works ended in British disaster.

This painting, based on eyewitness sketches of a British lieutenant, depicts the tenacious, but unsuccessful, defense of Lake Borgne by American gunboats on 14 December 1814. Securing the lake's passage enabled the British to make an amphibious assault on New Orleans. *U.S. Naval Academy Museum*

Evaluation

The war at sea proved a disappointment to both belligerents, but especially to the British. The profound shock of the American frigate victories in the first year of the war was compounded by the surprising success of American privateers, who operated even in the English Channel and the Irish Sea. Together, American public and private warships made more than 1,300 captures. The great size of the Royal Navy meant that trained seamen were spread thinly over the fleet and leadership was of uneven quality. In contrast, the regular American navy had excellent ships and was so small that the best officers could be appointed to command them. And as it was a navy of volunteers, the best seamen could be selected to man the few ships.

Yet, very quickly the British achieved dominance at sea. From the summer of 1813 Royal Navy ships commanded the waters around the United States, ruining American commerce and laying open the whole of the coastline to attack. Except on the northern lakes, the United States Navy was too small to engage a British squadron of any size. Its operations had to be restricted to seeking out single British frigates or smaller ships and raiding British commerce. It could neither challenge the strangling British blockade nor prevent harassment along the American coast and the burning of Washington.

Despite its overall failure, the Navy provided the nation moments of triumph otherwise few and far between. Its victories were romanticized and glorified. As a result of the War of 1812, the Navy became identified as the defender of national honor, economic interest, and individual freedom.

MJC

Through repetitive gunnery training of his crew, Captain Philip V. B. Broke, RN, secured victory for his Shannons over USS *Chesapeake* on 1 June 1813. *The Naval Chronicle, Vol. 33*

War of 1812: The Royal Navy

In 1812 the United Kingdom boasted a navy of some six hundred warships of all types (including first rate ships of the line of one hundred guns and more), manned by some 130,000 sailors, and an additional 250 warships under construction or fitting out. Only about two-dozen Royal Navy ships, however, were on the North American Station. Another fifty-seven warships were in the western hemisphere, on the Newfoundland, Leeward Islands, and Jamaica Stations. Most of the rest were enforcing the British blockade of the European continent or protecting Britain's far-flung empire.

The great size of the Royal Navy gave it a significant advantage over the miniscule Navy of the United States, of course. Size brought liabilities as well. The Royal Navy's demand for men meant that many of its crews were not professional sailors and that experienced seamen were spread thinly throughout the fleet. The large number of ships also meant that leadership was of uneven quality. The Royal Navy possessed excellent officers, such as Captain Philip B. V. Broke, who drilled his Shannons until they were expert gunners. But for every Philip Broke there were dozens of commanders who, given the absence of any credible threat at sea since 1805 when the British won naval dominance at the Battle of Trafalgar, set gunnery at naught.

The tradition of victory with which the Royal Navy entered war with the United States proved as much a liability as did the size of the fleet. The expectation of superiority fed complacency. Besides, having dominated the oceans since 1805, British officers and crews had much experience maintaining a blockade, but little recently of battle. The defeat and capture of three of their frigates in single-ship actions during the first six months of the war came as a rude shock to the Royal Navy and the British public. In response, and in recognition of the superior strength of American-built warships, the Admiralty instructed frigate commanders to avoid engaging singly against the American 44s.

The American declaration of war in June 1812 found Vice Admiral Herbert Sawyer in charge of the North American Station of twenty-three armed vessels scattered from Halifax to Bermuda. With these he was expected to cut off American commerce while protecting British merchantmen. Replacing Sawyer in September 1812, Admiral Sir John Borlase Warren repeatedly complained to the Admiralty that he had too few ships with which to keep the swarms of American privateers, as well as U.S. Navy ships, confined to port. The Admiralty reacted slowly in providing more ships, but did not hesitate to impose more responsibilities, including a rigorous blockade of American ports from the Delaware River south.

By the summer of 1813, the British Navy commanded American waters, tightened the blockade, and could support landings and raids almost at will. In 1813 and 1814 the British made amphibious raids on coastal towns of Maine, Massachusetts, and Connecticut, and up and down both shores of the Chesapeake Bay, where they terrorized towns and burned tobacco warehouses. U.S. warships, in contrast, no longer sortied freely, and several American frigates remained confined in port for most of the war.

Relieving Warren, Vice Admiral Sir Alexander F. I. Cochrane arrived on the North American Station in the spring of 1814, with orders, determination, and means (ships and men released from the European theater by Napoleon's capitulation) to pursue the war vigorously. Cochrane extended the blockade to include New England, offered runaway slaves asylum and opportunity to serve in the British army, and urged his commanders to retaliate for American actions in Canada. He directed Captain George Cockburn to intensify his joint campaign with the army in the Chesapeake region, in order to force America to recall soldiers from the Canadian border. On the Gulf Coast, Cochrane laid plans to use pirates, Indians, and runaway slaves to create a "fifth column" to divert American resources from combating British forces.

Despite their naval superiority, as well their increasing reinforcements of land troops especially after the exile of Napoleon to Elba, the British failed to win such victories as would have empowered them to dictate the terms of peace. Why was John Bull unable to force Brother Jonathan to his knees and make him cry "Uncle"? The interplay of three overarching factors worked against British victory—strategy, logistics, and timing.

The British never developed a war-winning strategy. The pinnacle of British success in the War of 1812 may have been the capture of Washington, D.C., the capital of the United States, and the burning of its public buildings. That British victory, however, had no strategic effect—Washington was neither an economic nor a logistical center. The temporary interruption of the operations of the federal government did nothing to disrupt the decentralized American war effort. British activities in the Chesapeake failed in their strategic goal, for they diverted no American soldiers or Sailors from the north, where the British hoped to win the war by capturing territory.

Since the British could not win the war by capturing vital centers, they must win either by capturing and holding significant extents of territory, or by destroying the American government's ability to continue paying for the war.

The best possibility of conquering territory was along the Canadian frontier. The British succeeded in capturing and holding Castine, Maine, but the region they held was not vital to the territorial integrity of the United States. The Royal Navy accepted responsibility for Canadian inland waters in 1813 and built substantial fleets on Lakes Ontario and Erie. Although the U.S. Navy's Oliver Hazard Perry supposedly worked a miracle in the wilderness by building his fleet on Lake Erie, the British faced much more daunting difficulties, having to transport all their

Rear Admiral George Cockburn, RN, coordinated several amphibious assaults against coastal areas in the Chesapeake Bay and off the North Carolina and Georgia coasts during the War of 1812.
Navy Art Collection, Naval Historical Center

building materials for Lake Erie from England and, not controlling Lake Ontario, over roads much more primitive than those south of the border. They lost the Battle of Lake Erie because they lost the shipbuilding race. Capture of their entire Erie fleet in the late summer of 1813 led to defeat on land at the Battle of the Thames and the end of plans for a permanent territory for Britain's Indian allies in the Northwest. Better supplied, the British Lake Ontario squadron held their American counterpart to a stalemate.

In the end, the Lake Champlain corridor proved the best hope for British victory. By the summer of 1814 a British army of ten thousand men had assembled in Canada ready to invade and capture U.S. territory. Unfortunately for them, their Lake Champlain squadron, long neglected by naval policy makers, lost the contest. With the American squadron in command of Lake Champlain and able to cut supply lines, the British army invading New York had to withdraw back into Canada. That withdrawal left the British with no claim to territorial concessions. Americans were also able to repulse British attacks at Baltimore, Maryland, and New Orleans, Louisiana. Of all the naval actions of the war, however, it was the Royal Navy's failure at Plattsburgh Bay that exerted the most profound influence on the peace treaty signed on Christmas Eve 1814.

The one remaining avenue of British victory would have been the financial ruin of the U.S. government. By the end of 1814 the Royal Navy's enforcement of the blockade had indeed brought the U.S. government near to bankruptcy. After some two decades of almost continual warfare with France, however, the British people were tired of war and the taxes that supported it. With peace at last established in Europe, they were not willing to continue long a war whose outcome was less crucial to them than their long conflict with Napoleon had been. The major joint campaigns in the Lake Champlain Valley, the Chesapeake, and the Gulf of Mexico were meant to bring the war to a quick and decisive conclusion. Those campaigns failed or came too late. Lacking the political will to wait for the American government's collapse, the British agreed to a peace without American concessions.

MJC

Shipboard Routine

Referring to his service in U.S. ship of the line *Columbus* from 1845 to 1848, Charles Nordhoff later recalled, "my station at quarters, or in time of battle, was as powder boy at gun No. 36, on the main gun-deck; my hammock number was six hundred and thirty nine; my ship's number, five hundred and seventy-four; and the number of my mess, twenty-six. Thus was the whole routine of my life on board this vessel laid out for me."

Every Sailor entered on the rolls of USS *Constitution,* as well as those of all other ships of the United States Navy during the age of sail, had his ship's number (found in the muster roll, and used by the purser to keep track of the moneys owing to, or owed by, each Sailor), his hammock number (painted or tacked on in tin in sequential order on the berth-deck beams from which the hammocks were slung by hooks), the number of his mess, his station in the ship for different evolutions, his division, and his quarters at the guns. And if a watch-stander and not an idler, he was assigned either to the larboard or to the starboard watch. These numbers and assignments to station, division, quarters, and watch regulated when and where a Sailor slept, ate, and performed his duties, and what those duties were. These numbers and assignments helped create order among some 450 men crowded into a relatively small space, guaranteed that each member of the crew knew his place and his duty, and assured that every order given had a particular man or set of men responsible for executing it. This arrangement, perfected in European navies over the preceding centuries, transformed a combination of men, ship, cannon, and gunpowder into an efficient and effective weapons system.

Take for instance the protagonist in Herman Melville's novel *White-Jacket, or The World in a Man-of-War,* for which the author drew on his experience in *Constitution*'s sister ship USS *United States.* White-Jacket belonged to the starboard watch, "and in this watch he was a main-top-man; that is, was stationed in the main-top, with a number of other seamen, always in readiness to execute any orders pertaining to the main-mast, from above the main-yard." White-Jacket's particular duty was to loose the main-royal sail. "Thus, when the order is given to loose the main-royal, White-Jacket flies to obey it, and no one but him." "In tacking ship, reefing top-sails, or 'coming to,'" as Melville explains, "every man of a frigate's five-hundred-strong, knows his special place, and is infallibly found there. He sees nothing else, attends to nothing else, and will stay there till grim death or an epaulette orders him away."

Time on board ship was divided by watches and reckoned by a number of strikes of the bell, from one through eight, each "bell" marking the passing of a half hour. Watches consisted of four-hour periods, beginning at midnight, except that the time between four and eight o'clock p.m. was divided into two periods, known as the first dogwatch and the second dogwatch. (See the "Bells and Watches" table below.) The device of the dogwatches meant that the larboard and starboard watches would alternate the midnight duty.

The chaplain, cook, purser, surgeon, boatswain, carpenter, gunner, sailmaker and their mates, the first lieutenant, and the captain and first lieutenant of Marines were known collectively as idlers, since they did not stand watches by night. Because their jobs kept them busy throughout the day, they were free to sleep through the night, except in emergencies when all hands were called. Watch-standers, in contrast, seldom slept four hours in a row, at midnight either being relieved or being roused from their sleep to go on watch.

Each member of a watch had a general station, as well as specific stations for various evolutions such as weighing anchor or tacking ship. The *forecastle* or *sheet anchor men*, the most skillful Sailors, handled the headsails and the bower anchors, with the best of them working in the foretop. Skilled *topmen* were stationed in the main and mizzen tops as well. The topmen of each station were further divided into quarter watches, allowing sets of topmen to relieve each other during a watch. Somewhat less skilled Sailors had their station on the quarterdeck, where they were known as the *afterguard*. The least skilled were stationed in the ship's waist and called *waisters*.

For battle stations, the crew was organized into five or six divisions, with a lieutenant in charge of each. The First, Second, and Third Divisions manned ten guns each, divided five on a side, on the gun deck; the Fourth Division manned the forecastle guns; and the Fifth, the quarterdeck guns. If the commanding officer desired, a sixth division would be dedicated to fighting any fires that might break out during combat.

A routine day at sea began at daylight, about a half hour before sunrise, when reveille would be called. After the seamen had rolled, tied, and stowed their hammocks and had holystoned the decks, breakfast was served, and the smoking lamp lit, allowing Sailors to smoke near the camboose in the galley. After breakfast, petty officers took up their crafts and the men on watch not occupied by more pressing duties turned to such routine work as picking oakum, repairing rigging, polishing brass, and blacking the guns. Midshipmen and ship's boys might attend classes. Twice a day, in the morning and in the afternoon, the men would be mustered at quarters. If any disciplinary actions were to be carried out, all hands would be called to witness punishment before noon. The one hot meal of the day was served at noon, when the day's first issue of grog also took place. The evening meal and the second issue of grog came at four p.m. After dinner the men would sweep the decks and then have time for relaxation. At sunset the hammocks were piped down. After lights out, the midshipman of the watch and the master-at-arms and his agents would assume that anyone else moving about below decks was up to no good.

The captain chose some days to be devoted to practice at the great guns, exercise with cutlasses and boarding pikes, and training in various sailing maneuvers. Each week had one or two days dedicated in part to washing and mending clothes. The men were expected to keep themselves clean and were allowed to shave

A Sailor's day ended on the berth deck (below the gun deck) with the ritual of unrolling, hanging, and then climbing into his hammock. *Old Ironsides, U.S. Frigate Constitution: An Essay in Sketches*

Bells and Watches

Bells	Night watch (a.m.)	Morning watch (a.m.)	Forenoon watch (a.m.)	Afternoon watch (p.m.)	1st dogwatch (p.m.)	2d dogwatch (p.m.)	Evening watch (p.m.)
1	12:30	4:30	8:30	12:30	4:30		8:30
2	1:00	5:00	9:00	1:00	5:00		9:00
3	1:30	5:30	9:30	1:30	5:30		9:30
4	2:00	6:00	10:00	2:00	6:00		10:00
5	2:30	6:30	10:30	2:30		6:30	10:30
6	3:00	7:00	11:00	3:00		7:00	11:00
7	3:30	7:30	11:30	3:30		7:30	11:30
8	4:00	8:00	noon	4:00		8:00	midnight

For instance, 10:00 a.m. is "four bells in the forenoon watch."

twice a week. The chaplain read prayers twice each day, attended by as many men as could be spared from duty, and on Sundays the men gathered on the spar deck to hear the chaplain deliver a sermon. Once a month the captain read aloud the Articles of War (See pp. 114–19, Act for the Better Government of the Navy, 23 April 1800) to the assembled crew.

For an example of the routine schedule established for one particular frigate, see the chart "Daily Routine at Sea in USS *Independence,* Drawn from General Orders of 1815," which follows.

Routine in port was similar to that at sea, with a few deviations. The morning gun at 8 a.m. announced to those on shore that the ship was open for business. In port, more time was devoted to ship maintenance than was at sea. Through the day, boats brought supplies and carried to and fro visitors doing business or paying social calls. At sunset the evening gun signaled that visitors should depart and that Sailors ashore were to return to the ship within half an hour. At dark a Marine guard was posted to prevent desertion and to detect any approaching boats.

Watches

Night watch:	midnight to 4 a.m.
Morning watch:	4 a.m. to 8 a.m.
Forenoon watch:	8 a.m. to noon.
Afternoon watch:	noon to 4 p.m.
1st dogwatch:	4 p.m. to 6 p.m.
2d dogwatch:	6 p.m. to 8 p.m.
Evening watch:	8 p.m. to midnight.

Daily Routine at Sea in USS *Independence*, Drawn from General Orders of 1815

7:30 a.m.	Hammocks piped up and decks swept
8:00 a.m.	Breakfast
9:00 a.m.	Provisions served out by messes; messes place their tallies on their portions
Noon	Dinner
1:00 p.m.	Decks swept
4:00 p.m.	Work areas cleared and decks swept
1 hour before sunset	Supper
Evening	Muster at quarters; hammocks piped down; lights out
10:00 p.m.	Wardroom lights out

Monday, Tuesday, and Friday mornings were dedicated to exercises with the great guns and small arms.

On Thursdays the men were allowed time to mend clothing. Early Saturday mornings, before the hammocks were piped up, the men washed their clothes in water heated by the cook.

After breakfast on Thursdays and Sundays, the crew washed and shaved, and changed their clothes, and then, on Sunday mornings, assembled for inspection in a muster by division and a general muster.

General Orders U.S.S. Independence, 1815 (Washington, DC: Naval Historical Foundation, 1969).

MJC

Iron Men for Wooden Ships: Recruiting the Fleet in the War of 1812

On the eve of the War of 1812, the U.S. Navy had an operational force that included fifteen ships of various classes and sixty-five gunboats. Once hostilities with Great Britain commenced, this small force increased rapidly through the recommissioning of previously laid up vessels, the capture and purchase of ships, and an aggressive shipbuilding program that launched craft ranging from fifty-foot barges to mighty 74-gun ships of the line. Readying this greatly enlarged fleet for service was a monumental task, demanding enormous amounts of money, materiel, and labor. It also required large numbers of men. One of the Navy Department's great challenges during the war was finding enough Sailors and Marines to man its ever-expanding, far-flung collection of warships.

Congress had regulated the number of men in naval service since 1794. By 1809, federal lawmakers had set the Navy's strength at nearly 5,600 officers and seamen. Three years later, Congress eliminated these restrictions in anticipation of the Navy's need to man a growing, wartime establishment. In addition to public warships this included Navy yards and stations spanning the American coastline from Portsmouth, New Hampshire, to New Orleans, Louisiana, as well as naval facilities built on the frontier shores of Lakes Erie, Ontario, and Champlain. By far the Navy's greatest manpower needs were afloat, especially among its largest classes of vessels: sloops of war, ships, and frigates. These warships carried complements of from 140 to 500 men. And while the more diminutive vessels on the Navy's lists—gunboats, galleys, and barges—had crews of 30 to 50 Sailors, their sheer numbers (130 in June 1814) necessitated a significant allocation of the Department's manpower resources. The steady loss of seamen through death, discharge, desertion, and enemy capture also contributed to the Navy's manpower concerns.

The Navy obtained its enlisted men in one of two ways: through drafts, or transfers of personnel from one ship to another, and the opening of naval rendezvous, or recruiting centers, in port cities and towns. Drafting afforded the easiest and fastest way to address manpower shortages. It also provided the means of giving useful employment to the crews of ships idled by the British blockade. Because service on Lakes Erie and Ontario proved so unpopular with most seamen, transfers of Sailors from Atlantic-based ships became the primary method of supplying crews to the squadrons on those stations. The quality of the men transferred, though, sometimes left much to be desired, prompting bitter complaints by squadron commanders such as Oliver H. Perry and Arthur Sinclair.

Most Navy vessels and stations met their manpower needs through the use of naval rendezvous, the 1812 equivalent of today's local Navy recruiting offices. Usually, they were opened at the direction of the secretary of the navy, whose instructions provided guidance on where the rendezvous was to be held, the number and class of seamen to be enrolled, and the wages and bounties to be paid out. Typically, these instructions were issued to ship commanders preparing their vessels for sea. Ships returning home from cruises shorthanded were likewise authorized to open rendezvous to enter replacement hands. Ideally, a ship opened a rendezvous in its homeport. But sometimes, local factors (the availability of seamen, competition with recruiters from other ships, etc.) compelled a ship's recruiting officers to seek men in more distant towns. Seamen's boarding houses were favored choices as headquarters for naval rendezvous. A large American flag displayed over the entranceway to one of these houses denoted a

Naval rendezvous, the nineteenth-century equivalent of today's Navy recruiting offices, were opened in major American seaports to enlist Sailors for the fleet. *Naval Historical Center photograph*

Edward Barry, John Robinson, John Ellis, John Smith, John Reynolds, Whitford Hudson, Hanse Andrews & Benjamin Rogers being duly appointed Seaman on board the Frigate United States John Barry do ſolemnly ſwear to bear true allegiance to the United States of America, and to ſerve them honeſtly and faithfully againſt all their enemies or oppoſers whomſoever; and to obſerve and obey the orders of the Preſident of the United States of America, and the orders of the officers appointed over me, and in all things to conform myſelf, to the rules and regulations which now are or hereafter may be directed, and to the articles of war which may be enacted by Congreſs, for the better government of the navy of the United States—and that I will ſupport the conſtitution of the United States.

Sworn before me, one of the Justices of the peace of Newcastle County in the District of Delaware on 6th of July, 1798. Joseph Tatlow

United States departed New Castle, Delaware, on its maiden voyage on the morning of 7 July 1798. As the date of this oath of allegiance attests, Navy ships often entered recruits up to the very day of their sailing. *Rare Book Room, Navy Department Library*

rendezvous headquarters. Other means by which Navy recruiters attracted Sailors' attention included word-of-mouth advertising, visits to mariner's haunts, playing music, and publishing notices in newspapers, as Captain Samuel Nicholson did when recruiting *Constitution*'s first crew in the spring of 1798.

The Navy expected recruiting officers to enlist none but active, healthy men, who possessed sufficient sailorly skill to serve in a man-of-war. Departmental policy discouraged and, at times, explicitly forbade the enlistment of blacks or foreign nationals, but Congress overturned such sanctions in "An Act for the Regulation of Seamen," passed in March 1813. Despite the passage of this law, racial prejudice continued to influence recruiting policy for the remainder of the war. As late as December 1814, John Rodgers was instructing his recruiting officers to withhold offers of bounties to black Sailors unless instructed to do otherwise.

Officers weeded out poor candidates for naval service by interviewing recruits on their background, work history, and experience at sea. Once satisfied as to a man's qualifications, the recruiting officer offered him a berth. On passing a physical exam administered by the ship's surgeon or surgeon's mate, the new recruit was sworn into service and signed shipping articles. The latter document specified the recruit's ship, his rating, and length of time to be served (most often, two years from the commencement of a cruise). At this point the new crewman received any bounties and advance wages authorized by the Department. As a protection against theft, these moneys were paid only to enlistees able to provide sureties (persons liable for the debt of another) for their appearance on board ship. Otherwise recruits received payment after mustering with the ship's company. Recruits who could not be immediately dispatched to their ship were placed temporarily in receiving vessels to prevent their desertion. An officer and guard of men escorted those destined for ships on distant stations.

The Navy employed a number of strategies to encourage seamen to join the service. One was to employ officers who enjoyed popular reputations within maritime communities. Joshua Barney sought to do this when he engaged Solomon Frazier of Maryland to recruit Sailors for the Chesapeake Flotilla. Frazier, wrote Barney, "is well known as a Character, perhaps the most popular among the seafaring men on the Eastern shore, of any man in Maryland." Additional inducements to recruits included shorter terms of enlistment (twelve months) and guarantees of non-draft status. These conditions were offered to Sailors in the New York and Chesapeake Flotillas. Money, however, proved the most powerful incentive to join the naval service. At some rendezvous the most experienced seamen received $20 bounty money plus a four-month's advance of wages ($48). These totals were even higher for men recruited for the Erie and Ontario squadrons after February 1814, when the Navy Department authorized a 25 percent increase in wages and bounties for those serving on the lakes.

At times the pressure to provide Sailors to the fleet was so strong that Navy recruiters were tempted to engage in wrongful or illegal practices to fill their quotas. Alcohol was used by less scrupulous officers to dupe grog-starved seamen into enlisting. Some officers enrolled men clearly unfit for duty at sea or shipped them at ratings above their skill level. Sailing Master Thomas N. Gautier was guilty of entering slaves, captured in British ships, as seamen in the gunboats under his command at Wilmington, North Carolina, and then pocketing their wages. The landlords of Sailors' boarding houses were notorious for committing the worst abuses associated with the recruiting service. They used threats of jailing to compel their Sailor-tenants to enlist in the Navy, and then claimed the recruit's bounty and advance wages in compensation for unpaid debts.

Despite the best efforts of Navy administrators and rendezvous officers, the fleet remained chronically short of seamen throughout the war. There were a number of factors that contributed to this failure. One was the strong competition for seamen faced by the Navy from privateers and the Army. Seamen were attracted to privateers for the greater possibilities such ships offered for prize money; while service in the Army was rewarded with a significantly higher bounty ($124 plus 150 acres of land) than that offered by the Navy. After the lifting of the embargo in April 1814, Navy recruiters also found themselves competing with the merchant marine for seamen. Many rendezvous officers were dismayed as well to find the Navy in competition with itself, when as many as four ships at a time vied for seamen in a single port. The unpopularity of service on Lakes Erie and Ontario, the site of the Navy's most important operations, also did much to depress recruiting totals. Perhaps the most decisive factor in crippling the Navy's recruiting efforts was the lack of money. The Navy never had adequate funds to offer the kinds of bounties and wages necessary to fill the fleet's billets. Moreover, shortages of money regularly brought recruiting to a halt or prevented the reenlistment of Sailors ready to receive their discharge.

The most damaging consequence of the Navy's manpower shortage was on its operational readiness. Increasingly, as the war went on, American cruisers were unable to challenge the enemy's blockade and get to sea because they lacked adequate manning. Naval commanders charged with the nation's coastal defense similarly experienced a decreasing ability to execute their mission as they witnessed their flotillas depleted by drafts, death, and desertion. Though the offensive and defensive capabilities of the American fleet had been severely blunted by war's end, it remains impressive that some of the Navy's greatest ship-to-ship victories of the war were achieved in 1815. When combined with its earlier combat successes, the Navy's underfunded, undermanned performance during the War of 1812 demonstrated to the world the potentialities of a fleet manned by free Sailors.

CEB

Gun Drill in Old Ironsides

The reason for American success in ship-to-ship actions in the War of 1812 was no mystery to Enoch Wines, a schoolmaster in the frigate *Constellation* during its Mediterranean cruise of 1829–1831. The "brilliant victories" of ships such as *Constitution* and *United States*, he declared, were owing to the Navy's superiority in gunnery. As Wines contemplated his service's future, he had good cause to feel encouraged about its prospects as an able combat force. "Our naval officers," he observed,

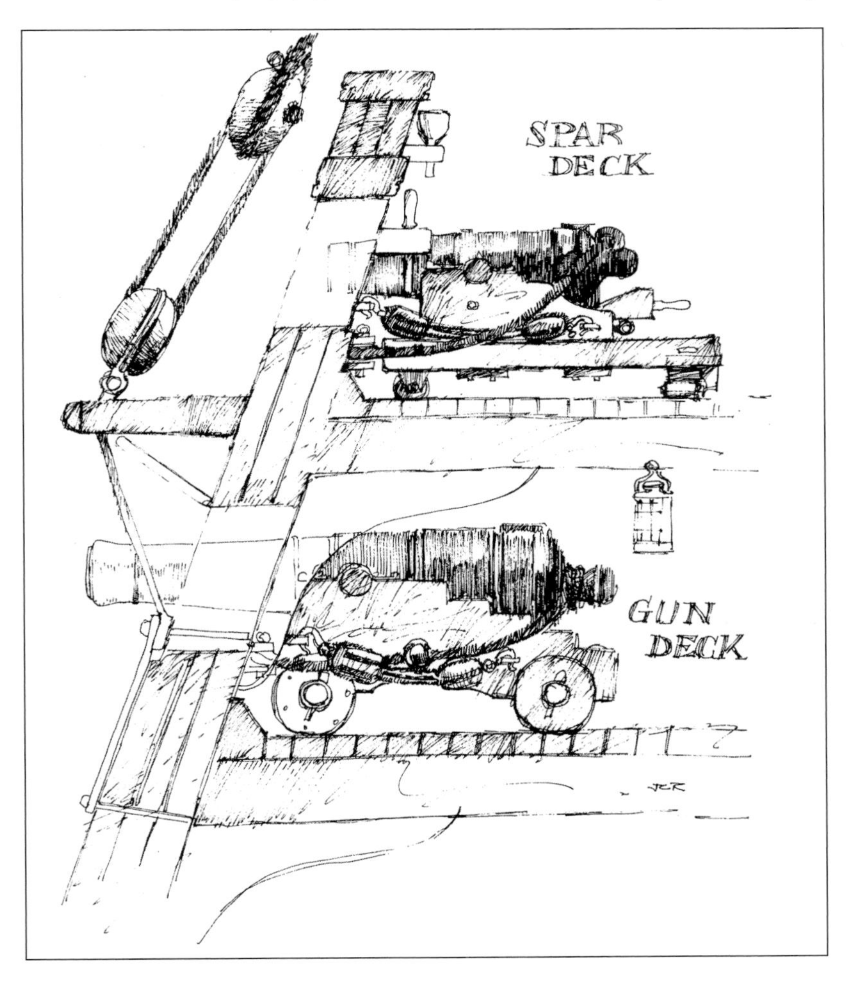

Cutaway view illustrating the relative size and positioning of *Constitution*'s cannon. The frigate carried carronades on its spar deck and long guns on its main or gun deck.
Old Ironsides, U.S. Frigate Constitution: An Essay in Sketches

Regular gunnery drill, including live fire target practice, contributed to *Constitution*'s dominance in ship-to-ship actions during the War of 1812. *Naval Historical Center photograph*

"are unrelenting in their exertions to render our seamen expert and ready" in the art of gunnery. Writing at nearly the same time as Wines, British Major General Sir Howard Douglas, author of *A Treatise on Naval Gunnery,* offered a similar analysis of Anglo-American sea fights in the War of 1812. Douglas attributed American success not only to better-executed battle tactics by Yankee commanders but also to more skillful gunnery by U.S. crews. If England was to avoid similar reversals in the future, he argued, then the Royal Navy must recognize "the absolute necessity of training to expert practice, master-gunners, their crews, and captains of guns."

The conclusions of Wines and Douglas underscore the central fact of combat at sea in the age of sail—the decisiveness of gunnery. Neither carefully chosen tactics, nor skillful ship handling, nor even absolute superiority in number and weight of guns could guarantee victory, if a ship's cannons were not well served. It was the ship that fired more quickly and accurately than its opponent that most often secured victory in battle. In the U.S. Navy there is ample evidence that its commanders took seriously the necessity of drilling their crews at gunnery. As the American author of an 1813 treatise on seamanship counseled, "This most principal part of naval discipline can never too much engross the care of a commander, since the honor of his ship is so nearly allied to it, as the primitive step to victory, or a hard contested action." Ship captains ignored such advice at their peril.

Navy Department regulations in 1812 called on commanders "to frequently exercise the ship's company in the use of the great guns and small arms." While the regularity of gun drill in any ship was left to the discretion of its captain, there were circumstances that sometimes affected how often Sailors trained at their guns, as, for example, when foul or extreme weather made safe exercise of the guns impossible. A crew's experience, or inexperience, also dictated the frequency of gunnery drill. Preparing for his first wartime cruise in *Constitution,* Isaac Hull found himself in the unenviable position of having a large number of "greenhorns" among his crew, many of whom had never served in a man-of-war before. The Connecticut-born captain exercised his men at the great guns up to two hours daily over several weeks to infuse them with enough confidence and competence to face the enemy in battle. Similarly, Charles Stewart exercised his crew at the guns almost daily when compelled to press *Constitution's* Marines into service as gunners after more experienced hands were sent to man the prizes *Cyane* and *Levant.* William Bainbridge, despite inheriting a veteran crew from Hull in the fall of 1812, exercised Old Ironsides' men at the guns three times weekly, a practice he carried over to his 1815 command, the 74-gun ship *Independence.* Besides drilling frequently at sea, American tars also developed their gunnery skills in port. According to one participant, the crew of the U.S. brig *Siren,* anchored in Boston Harbor in the fall of 1813, was "frequently exercised in the various evolutions of a sea-fight" including drill at the brig's great guns while awaiting orders for sea.

Preparatory to any gun drill, a ship's company first had to be assigned to their stations in battle. This duty was performed by the ship's first lieutenant who assigned each man his quarters in combat. The document listing each Sailor's assigned station was called a quarter bill and was posted below decks for all hands to consult. The number of men assigned to each gun depended on the class and caliber of gun being served. Generally carronades required fewer men to serve them than did long guns. Also, the heavier the shot thrown, the larger the crew needed to work the cannon firing it. In the War of 1812, eight to nine men worked *Constitution*'s 32-pounder carronades while nine to fourteen men served its 24-pounder long guns. Each crew was composed of gun captains, tacklemen, swabbers, loaders, and powder passers to clean, load, move, aim, and fire their piece. In the event that the ship's maneuvering necessitated shifting fire from one side of the vessel to the other, each gun crew moved to the cannon on the opposite side of the deck. If circumstances required it, each gun crew could be divided in half so that both starboard and port batteries might be fought simultaneously.

Guns and their crews were grouped in divisions (usually of ten guns, five to a side) under the direction of a lieutenant. Each division was further subdivided and placed in the charge of a midshipman. William Bainbridge organized *Constitution*'s guns in five divisions: forecastle and quarterdeck on the spar deck; and first, second, and third divisions on the gun deck. When called to exercise at the great guns, the gunner and his mates instructed the men at their duties. This was performed under the watchful eye of the subdivision and division commanders. There was no standard set of commands for exercising the great guns in 1812. But an 1813 American text, reproduced below, gives twenty individual commands for that exercise. Undoubtedly, captains tailored the wording and number of commands to promote the speedy and efficient handling of batteries.

While gunnery drill in the U.S. Navy was often performed without ammunition, it is clear that American ship commanders also had their men fire at targets with live ammunition. No doubt some captains were reluctant to expend precious powder and shot in gunnery exercise when it might be needed later in battle. However the benefits of live fire target practice in promoting efficient and accurate gunnery were too obvious for some American commanders to decline adopting. According to Moses Smith, a Sailor in *Constitution,* the American frigate's victory over *Guerriere* was a result of superior gunnery skills honed during weeks of target "shooting at hogsheads in the Chesapeake." Ned Meyers who served in USS *Scourge* on Lake Ontario described how regular practice in firing at targets on the lake enabled the men in his tiny schooner to become "rather expert cannoneers."

The skill of American Sailors at handling their guns and giving cool, well-directed fire in ship-to-ship actions resulted in a number of remarkable victories over the British during the War of 1812. Had more officers in the Royal Navy trained their crews to the high state of discipline in gunnery exhibited by the crew of Captain Philip V. B. Broke, captor of USS *Chesapeake,* then the American navy might have had less to boast about at war's end.

Gunnery Drill Commands, 1813

1st...*Silence!*

The men are to keep a strict silence, and pay an implicit obedience to their officer's orders.

2d...*Cast loose your guns!*

The trainers are to cast loose the gun, and hook the train-tackle, the swabbers at the same time will either coil down, or choke, the side-tackle falls, as the motion of the ship may render it expedient.

3d...*Take out your Tompions!*

The powderman is to remove the Tompion from the gun, with an iron and mallet, for that purpose, and place it out of the way, where it may be found after exercise.

4th...*Take off your aprons!*

The trainer who stands abaft the gun, is to take off the Apron, and the one, who stands forward, should have the powder-horn ready to give to the captain of the gun at the next word.

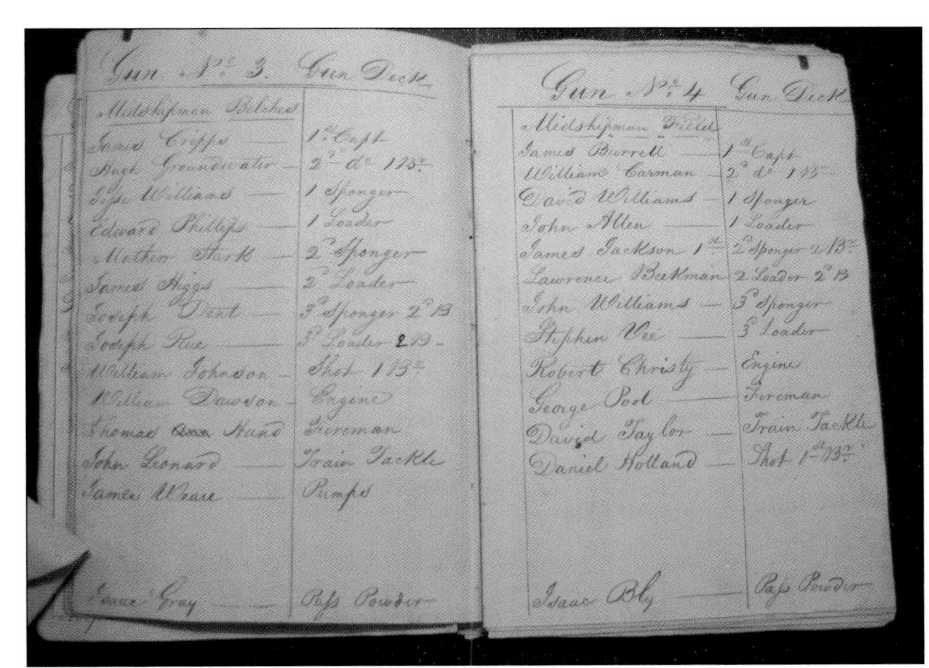

Gun No. 3. Gun Deck

Midshipman Belcher	
James Cripps	1st Capt
Hugh Groundwater	2d do 1 B
Jesse Williams	1 Sponger
Edward Phillips	1 Loader
Mathew Stark	2 Sponger
James Higgs	2 Loader
Joseph Dent	3d Sponger 2d B
Joseph Rue	3d Loader 2 B
William Johnson	Shot 1 B
William Dawson	Engine
Thomas Ann Hand	Fireman
John Leonard	Train Tackle
James Weare	Pumps
Isaac Gray	Pass Powder

Gun No. 4 Gun Deck

Midshipman Field	
James Barrett	1st Capt
William Carman	2d do 1 B
David Williams	1 Sponger
John Allen	1 Loader
James Jackson 1st	2 Sponger 2 B
Lawrence Beekman	2 Loader 2 B
John Williams	3 Sponger
Stephen Vee	3 Loader
Robert Christy	Engine
George Pool	Fireman
David Taylor	Train Tackle
Daniel Holland	Shot 1st B
Isaac Bly	Pass Powder

Constitution's 1812 quarter bill, listing the names and assignments of men stationed at two of the frigate's thirty gun deck cannon. *Record Group 45, Subject File, Entry 502, National Archives and Records Administration*

5th...*Handle your Powder-horns!*

The captain of the gun, receives the powder-horn from the Trainer, pricks the cartridge to know that it is home, and wipes the loose particles of powder off the priming wire, on the back of his hand to learn if the charge is dry; if wet, he will report the same to the officer of his division, who will order it to be drawn, and the gun re-loaded.

6th...*Prime your Guns!*

The captain of the gun half cocks the lock, places a tube on the vent, tearing first the paper from its face, and covers it, as well as the pan, with loose priming, which he should be careful in bruising with his powder-horn.

7th...*Return your Powder-horns!*

The captain of the gun returns the powder-horn to the trainer who removes it to a place of safety.

8th...*Lay on your Aprons!*

The trainer who took the Apron off, lays it on again.

9th...*Point your guns to the object.*

2...Points abaft the beam, &c. &c.

The trainers handle their crows and handspikes and by the direction of the Captain, train the gun to the object.

10th...*Stand by and take off your Aprons!*

The Apron will be taken off as before, and the trainers will train the gun to the eye of the captain, who will take as exact aim as possible, either at the horizon, a ship in company, or a cask thrown overboard for that intent, for he should remember that in action one shot to do execution, is better than a whole broadside thrown hastily away. The swabber who stands forward of the gun, will have the powder-horn ready to renew the priming should it flash in the pan, and the one who stands aft the linstock ready to give the captain should the gun hang or miss-fire.

11th...*Fire!*

The captain of the gun, sure of his mark, is to discharge the piece, and instantly stop the vent with a twisted piece of oakum made for that purpose. The swabbers swab round the gun, and the sponger prepares to sponge it.

12th...*Run in your Guns!*

The men clap on the train-tackle-fall and run in the gun. N.B. Most guns will recoil in of themselves, but seldom or ever far enough, or in a position for loading.

13th...*Sponge your Guns!*

The sponger is to sponge the gun; he should be careful in rubbing well round the chamber, to extinguish any loose particles of fire that may remain, as well as to knock the same off of the sponge on the cell of the port. N.B. Every third round the powderman is to worm the gun.

14th...*Load with Cartridge!*

The powderman receives the cartridge from the swabber, who stands aft, and shoves it into the gun as far as his arm will permit, being careful to place it arse foremost and seam downwards. The boy receives the empty powder-box from the swabber, and returns it to the hatchway for a full one.

15th...*Ram home your Cartridge!*

The sponger rams home the cartridge, and the captain of the gun pricks it with his priming wire at the vent to know when it is home.

16th...*Shot your Guns!*

The swabber who stands forward, hands the shot to the powderman who puts it into the gun. The captains of the guns may load with round and grape, but should never fire two round shot together, as it endangers the gun.

17th...*Wad to your Shot!*

The swabber who stands aft hands the wad to the powderman who puts it into the gun.

18th...*Ram home Wad and Shot!*

The sponger rams home both and returns the sponge.

19th...*Man your side-tackle-falls!*

The men on each side of the gun man the side-tackle-falls and prepare to run out the gun. The trainers, if they are exercising the weather guns, unhook the train tackle, but if the lee ones, they ease the gun out.

20th...*Run out your Guns!*

The gun is run out, the captain taking particular care to overhaul the breeching that it should not choke the trunks and retard the progress of the gun.

Seamanship both in Theory and Practice (New York: Edmund M. Blunt, 1813), 248–49.

CEB

Doctors Afloat: Medical Care in *Constitution* in the War of 1812

On the afternoon of 14 August 1812, less than two weeks out of Boston, a loud cry rang out from *Constitution*'s masthead: "Man overboard!" One of the frigate's crew, John Linsey, had tumbled from the main chains into the sea. Orders were quickly given to bring the ship to, lower the stern boat, and save the fallen Sailor. When the rescue party reached Linsey, some two hundred yards astern of the ship, he was still alive, though exhausted and drained of color. Linsey's survival owed to good fortune and the fact that he was an excellent swimmer (a skill some *Constitution* men surprisingly did not possess). Reflecting in his journal on Linsey's near-death experience, *Constitution*'s surgeon, Amos Evans, observed: "The tenure of a Sailor's existence is certainly more precarious than any other man's, a soldier's not excepted."

With injury, sickness, and death common features of naval service in the War of 1812, the life of an American Sailor, as Surgeon Evans noted, was a perilous one indeed. While Navy doctors could not eliminate all hazards of the sea service, they could minimize some. In caring for those fallen from illness, accident, or combat, and, in overseeing the health of seamen ashore and afloat, Navy surgeons played a vital role in maintaining the morale and operational readiness of the U.S. fleet.

At the outset of the War of 1812, the Navy's medical corps consisted of forty-six surgeons and surgeon's mates attached to the fleet's ships, yards, and stations. By war's end this number had nearly tripled to 128 officers. The majority of these men had received their medical training as apprentices to physicians. However, a significant number held medical diplomas or had received university-level instruction at such institutions as Columbia College and the Pennsylvania Hospital, while still others had attended European medical schools. Overall, the doctors serving in the wartime corps were young and had little experience in treating combat injuries. Amos Evans, for example, was twenty-six at the time of Old Ironsides' engagements with *Guerriere* and *Java,* while his two assistants, Surgeon's Mates John D. Armstrong and Donaldson Yeates, had received their commissions only a month before the commencement of hostilities. Until the establishment of the Bureau of Medicine and Surgery in 1842, the secretary of the navy acted as the medical corps's head and chief administrator.

Constitution was authorized to carry three medical officers, a surgeon and two surgeon's mates. Their primary duty stations aboard the frigate were sick bay and the cockpit. Sick bay was where the surgeon and his mates administered the day-to-day care of the ship's ailing and injured. Crewmen disabled by accident or illness were also berthed there. *Constitution*'s sick bay was situated in the extreme forward section of the berth deck occupying a triangular space in the bows of the ship. It measured 30' at its widest, 18' along its centerline, with 6' of overhead space. It was set off from the rest of the berth deck by partitions and furnished with wooden platform beds with raised sides, a chair and desk for the medical staff, instruments, and medicines. Being two decks below the spar deck, the sick bay was poorly lit and ventilated.

The second duty area of the ship's medical department was the cockpit, located on the orlop deck between the main and mizzenmasts. This was where the surgeon and his mates treated battle casualties. *Constitution*'s cockpit measured 13' x 8', with an overhead space of 4' 10". Though lack of space, light, and fresh air made the cockpit a difficult space to work in, its location (below the waterline) rendered it a safer and more stable place to perform surgery than the sick bay. In smaller American warships, such as brigs and schooners, the officer's wardroom served as the surgery in lieu of a cockpit.

The surgeon and his mates cared for injured and ailing Sailors in the ship's sick bay. *Old Ironsides, U.S. Frigate Constitution: An Essay in Sketches*

Tools of the trade for *Constitution*'s medical officers. *Naval Historical Center photograph*

Constitution's surgeon performed a range of duties in overseeing the health of the ship's company. Each morning he held sick call to treat those seeking medical attention. Twice a day he tended to the patients berthed in sick bay. He informed the captain daily on the state and numbers of sick on board and advised him on any necessary measures to improve the healthfulness and well-being of the crew. He also had charge of the ship's medical stores (special foods, medicines, supplies, and instruments) and was responsible for keeping accounts of their purchase, receipt, and expenditure. In addition, he was required to keep a written record of his practice, detailing each patient's diagnosis and treatment history. Finally, the surgeon supervised the surgeon's mates to whom much of the daily care of the sick devolved. These latter officers assisted the surgeon in compounding prescriptions, dispensing medicines, changing dressings, and daily record keeping.

The ringing of a bell on the gun deck at 9:00 a.m. signaled sick call in *Constitution,* the beginning of the workday for the ship's surgeon and his mates. This was the established time for crew members to have their medical needs addressed, though men who suddenly fell ill, or suffered serious accident, were free to report to the surgeon at any time. Among the more common maladies surgeons treated aboard ship were respiratory complaints (colds, flu, pneumonia), digestive and intestinal disorders (diarrhea, dysentery), venereal diseases (gonorrhea, syphilis), fevers (malaria, yellow fever, typhoid), and scurvy. Surgeons also applied their healing talents to the victims of shipboard accidents including those with burns, cuts, sprains, contusions, concussions, bone fractures, and gunshot wounds. After examining and treating these cases, and the invalids confined to sick bay, the surgeon forwarded a list of the sick and their ailments to the captain for review. The sick list was then posted in the drawer of the ship's binnacle for inspection by the officers of the watch. The Sailors whose names appeared on this list were excused from doing duty. Except for the surgeon's afternoon rounds in sick bay at 4:00 p.m., the care of the sick fell largely to the surgeon's mates and their enlisted assistants known as loblolly boys. These male nurses fed, bathed, shaved, and saw to the lavatory needs of their bedridden shipmates. Gravely ill patients were discharged to shore hospitals when the frigate was in port.

When *Constitution* went into battle, its medical staff retired to the cockpit and prepared to receive the wounded. They laid out surgical instruments (saws, knives, probes, tourniquets), supplies (bandages, compresses, needles, thread), medicines (salves, ointments, painkillers), erected a crude operating table out of boards and mess chests, and spread sand on the deck to prevent slippage in spilled blood. As casualties occurred, they were carried down the hatchways to the cockpit. The wounds of the fallen ranged in severity from minor burns, cuts, and punctures to massive traumas to head, limbs, and vital organs. The surgeon and his mates treated the most grievously hurt first. A patient's chances of survival depended on the type of wound sustained. Shrapnel wounds to the chest and belly, for example, were generally regarded as inoperable. Arms and legs shattered by projectiles, on the other hand, were not, and were dealt with by amputation, an operation performed without benefit of anesthetics or antiseptics. Despite the best efforts of the frigate's medical officers some wounded never recovered, dying from complications (tissue loss, shock, infection) related to their injuries or surgeries. In *Constitution*'s three wartime engagements eight of its forty-six wounded officers and men died from their wounds.

Two principal theories informed naval physicians' understanding of the causes and treatment of disease in the early nineteenth century. The first, humoralism, dated to classical antiquity, and argued that an individual's health was regulated by four bodily humors (blood, phlegm, yellow bile, and black bile), each of which was associated with several special characteristics (moisture, dryness, heat, and cold). As long as these four humors remained in balance, the body remained healthy; when they did not, illness occurred. The second prevailing theory, solidism, had achieved acceptance by 1700, and held that a person's health was linked to the flow of the four humors through one's veins, arteries, and nerves. Disturbances in the strength, tone, and elasticity of these vascular and nervous tissues disrupted the proper flow of the body's humors, resulting in sickness. The concepts of these two complementary theories led physicians to diagnose and treat ailments based on a patient's symptoms. These symptoms provided clues to the type of physiological imbalance occurring in the afflicted person's body. Correcting the balance with the proper medicines restored the body to health. Stated another way, physicians treated a patient's symptoms, not his disease.

Naval surgeons had upwards of two hundred drugs in their medicine chests to dispense to the sick onboard ship. Depending on the symptoms being treated, they could prescribe medications from one or more of the following categories: emetics (to induce vomiting), cathartics (to stimulate the bowels), diuretics (to promote urination), diaphoretics (to produce sweating), narcotics (to provide pain relief), and tonics (to strengthen the body). In order to restore a patient's physiological balance, drugs were chosen that caused countervailing effects to the symptoms being exhibited. Thus, to treat a rapid pulse and elevated body temperature, drugs were administered that "cooled" the blood and "calmed" the rapid action of the veins and arteries. It is important to note that as a patient's symptoms changed through the course of his illness, so too did the drugs prescribed for him by the surgeon. In other words, there was no specific drug regimen for any illness. Instead, a surgeon tailored his treatment plan to restore the unique internal balances of each patient.

Surgeons and their mates were not the only ship's officers charged with overseeing a crew's health. Navy regulations specifically charged each vessel's commander with establishing rules to promote the cleanliness and health of his men. Such rules typically made tasks relating to shipboard hygiene (sweeping and scrubbing decks, washing clothes and bedding, cleaning and inspecting mess chests, removing trash and the like) part of the crew's daily and weekly routine. Other health measures captains instituted for their crews' well-being included improving ventilation below decks, fumigating the ship's hold to combat vermin, and ordering the purchase of fresh foods and vegetables for the men's mess tables. By making the living and working spaces aboard ship as clean and comfortable as possible, Navy captains made a significant contribution to the health and morale of the men under their command.

While wartime medical records for *Constitution* are incomplete, they do suggest that Old Ironsides never experienced the devastating incidence of sickness and disease that sometimes overwhelmed U.S. vessels on other stations, particularly those of the Erie and Ontario Squadrons. The numbers of men appearing on the frigate's sick list during various cruises ranged from eleven to forty-five in number or, from 2 to 10 percent of the crew; and only twenty seamen ever became ill enough to require discharge to hospitals ashore. Of the fifty-two deaths recorded in ship's papers, twenty-seven were combat related, five were accidental, and twenty were owing to sickness. Although these are admirable figures, most of *Constitution*'s sick recovered in spite of, rather than because of, the ministrations of the ship's doctors. It was in the treatment of battle casualties, however, that Old Ironsides' medical officers proved indispensable, for without their skillful surgical efforts most of the frigate's wounded would never have recovered.

CEB

Battle Tactics and Strategies

During the age of fighting sail and muzzle-loaded ordnance, several factors influenced the tactics employed when two warships engaged in combat on the open sea. Some of the more significant of these were the time of day, the state of the sea, the weather, the objectives of the combatants, the relative strengths of the opposing ships in men and guns, and the confidence of the commanders in their crews' fighting abilities and morale. But the factors that principally affected tactics of ship duels were three: (1) the structure of ships of war; (2) the range of the guns; and (3) the dependence for movement on the direction and force of the wind. During its active wartime service, USS *Constitution* cruised singly or in small squadrons, and never as a frigate attached to a fleet of line-of-battle ships. Therefore, this discussion will disregard the large body of theory on fleet tactics and focus on the tactics of ship-to-ship duels.

The Structure of Ships of War

By the eighteenth century, when *Constitution* was built, warships were constructed with heavy timber frames along the sides. These timbers, overlaid with thick planking, enabled warships to absorb much of the impact of enemy broadsides. *Constitution* and its sister ships boasted unusually strong sides for frigates, with timbers spaced as closely together as those of ships of the line. From this thick shielding came *Constitution*'s nickname, Old Ironsides. A warship's bow and stern, in contrast, lacking these heavy timbers, were relatively weak and vulnerable to penetrating shot. Solid shot

In possibly the most decisive action of the War of 1812, Commodore Thomas Macdonough, USN, used the geographic configuration of Plattsburgh Bay on 11 September 1814 to his defensive advantage, enabling his ships to hurl the first devastating blows against the approaching British squadron. *Navy Art Collection, Naval Historical Center*

piercing the stern could sweep the gun deck, decimating the crew and dismounting cannon. Similarly, all manner of shot fired at the bow or stern of a ship would range the length of the weather deck, giving the shot much better chance of hitting something than if fired across the deck from side to side.

Commanders sought to exploit the vulnerabilities of an enemy's ship by sailing at right angles athwart the enemy's bow or stern, a maneuver referred to as *crossing the T*, in order to rake the rival ship lengthwise, while striving to deny the opposing force opportunities to do the same to them. This maneuver could cause havoc, as when, during its duel with *Constitution, Java* crossed astern of the American frigate and unleashed a broadside that destroyed *Constitution*'s wheel, felled all four helmsmen, and gave Captain William Bainbridge his second wound of the battle.

A tactical challenge confronting every commander was to bring his vessel's guns into range of the enemy without exposing his own ship to the enemy's raking fire. The quickest way to engage the enemy was to bear down at a right angle and then to luff up to exchange broadsides. This method minimized the duration of the approach but exposed one's ship to the enemy's raking fire during the whole approach while depriving one's own ship of the power to reply with its own guns, apart from one or two bow guns. The opening moments of the Battle of Lake Champlain illustrate the hazard run in adopting this tactic. The American squadron had anchored in Plattsburgh Bay where the British squadron could engage it only by first sailing toward it at right angles. The first shot fired by the American flagship, USS *Saratoga,* from a single 24-pounder long gun, raked the British flagship, HMS *Confiance,* from stem to stern with devastating effect. During the approach, *Confiance* lost both of its bow anchors to *Saratoga*'s broadsides. Unable to reach the head of the American line without taking unacceptable losses, the British commander was compelled to alter his tactical plan and come to, opposite *Saratoga*.

To approach the enemy at an oblique angle until the two ships were alongside each other more or less equalized the opportunities of the opposing ships to exchange fire during the approach but also gave the enemy a better chance to avoid the engagement if he so desired.

Ships approaching from opposite directions could come alongside each other, holding fire until the foremost broadside guns crossed; or either ship could yaw during the approach to fire occasional broadsides. In this kind of approach, each commander would be especially wary to prevent the other ship from crossing the T by turning to cross bow or stern.

In a stern chase, yawing by the chase in order to fire broadsides in hopes of injuring the pursuer's sailing allowed the chasing ship to draw closer, whereas yawing to fire broadsides by the pursuer gave the chase a chance to widen the distance between the ships. During USS *President*'s chase of HMS *Belvidera,* the British commander criticized the American for this latter procedure, "yawing repeatedly and giving starboard and larboard Broadsides, when it was fully in his power to have run up alongside the *Belvidera*."

The Range of the Guns

The relative mix of long guns and short-range carronades strongly influenced a commanding officers' decision as to the distance at which to fight an engagement. Other considerations being equal, a commander who knew he had an advantage in long guns would try to fight the battle at a distance, staying out of range of the enemy's powerful carronades and using his long guns to disable the enemy ship's ability to maneuver by damaging its masts and rigging. Similarly, a commander of a ship armed mainly with carronades would seek to close with the enemy as quickly as possible in order to avoid being disabled outside the range of his main battery, and to use the smashing power of his guns to batter the opponent into submission at close quarters. Like their British counterparts, American frigates, with a few notable exceptions, carried a balanced mix of long guns and carronades. Whereas British frigates, however, carried 18-pounder long guns, Old Ironsides and its sister frigates mounted 24-pounders, giving the latter a decisive advantage in weight of metal thrown in broadside. The advantage in long-gun strength and in overall weight of broadside gave the Americans the tactically sound options of fighting either at long range, as Stephen Decatur in *United States* elected to do in fighting HMS *Macedonian,* or in a close-up slugfest, as Isaac Hull in *Constitution* did in fighting *Guerriere*.

The Dependence for Movement on the Direction and Force of the Wind

Frigates, being square-rigged vessels, could not sail against the wind. Thus, a frigate to the windward of another square-rigged warship had the option of choosing the moment to engage or of avoiding combat altogether. Tacticians referred to keeping to the windward as *holding the weather gauge*. With ships engaged broadside to broadside, the sails of the windward vessel could becalm those of the leeward vessel, eliminating the latter vessel's ability to maneuver. Being to the windward, however, could prove a serious detriment in heavy winds. Broadside to broadside in heavy winds, the

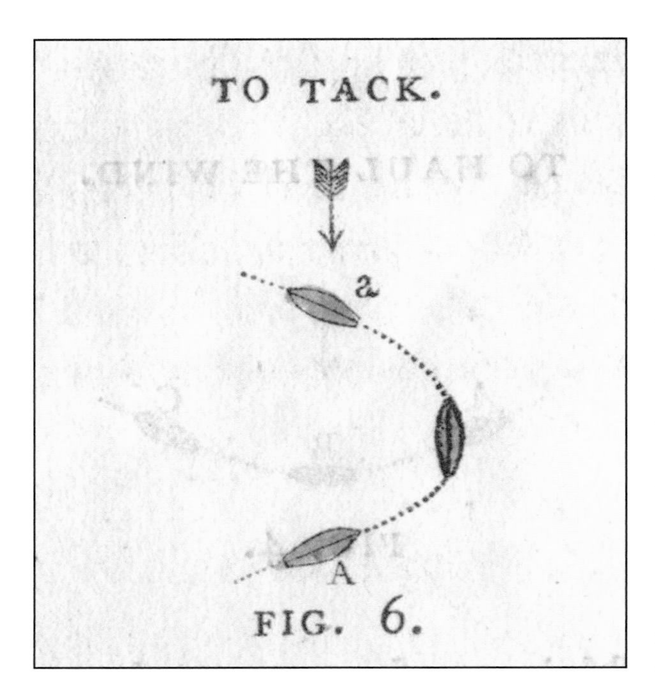

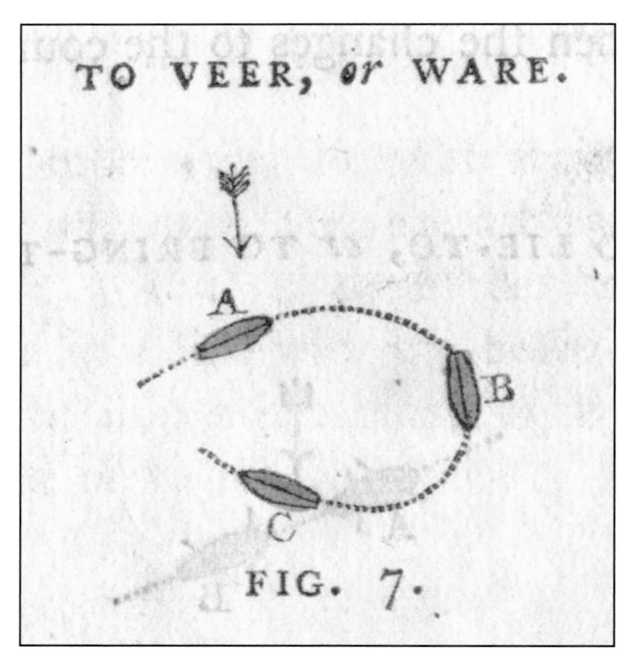

A ship working to windward was said to sail on either the starboard or port (larboard) tack depending on which of its sides met the wind first. Two maneuvers ship captains used to alter the tack, or course, of their vessel, were tacking and wearing or veering. In tacking, a ship changed its course from one tack to another by turning its head to the wind. In wearing, a ship altered course by turning its stern to windward. *A System of Naval Tactics, 1797*

ship to the windward could find its engaged gun deck depressed to a point so low that the guns could not fire, and its weather deck exposed and unprotected against the enemy's broadsides. In such a situation, being to the leeward was not without its disadvantages as well. While the ship to the leeward had its weather deck tilted away and protected from enemy fire, its hull below the waterline was exposed to the risk of being pierced.

Major factors on the open sea, the direction and force of the wind were vital in enclosed waters. A ship unable to round a headland without numerous tacks or embayed with the shore to leeward lost maneuverability much as if its sails and rigging had been disabled.

Boarding and Other Means of Securing Victory

A ship-to-ship duel could end in any of a number of ways. One ship might break off the engagement and escape before being captured, as the French frigate *Vengeance* did during its fight with USS *Constellation* in the Quasi-War. An engagement could end with one of the ships blowing up, as was the case of the Continental Navy frigate *Randolph* when it fought HMS *Yarmouth* in 1778. Such a victory could seldom be attributed to superior tactics rather than to chance. When a warship was so badly damaged that it lost ability to maneuver, the opposing ship could force surrender by taking a raking position athwart the bow or stern. The commander of H.M. brig *Avon* refused to surrender until U.S. sloop of war *Wasp* had reduced it to a sinking condition. A most effective way to end a ship-to-ship duel was by boarding.

In training for battle, next to importance to exercise at the great guns was small arms practice in preparation for boarding and repelling boarders. To board, a ship ran up alongside the enemy and lashed the two ships with grappling hooks and lines to insure that they would not separate and leave a boarding party outnumbered on board the enemy. With small arms, Marines in the tops would clear a space on the enemy's deck for the boarding party to cross and occupy. Led by officers, boarders, armed with cutlasses, pikes, and axes, rushed across, using every means to instill panic in the enemy. Columns of boarders would charge down each gangway attempting to gain the quarterdeck and compel the commanding officer into surrendering. In the event of a reversal, Marines provided covering fire to support their retreating comrades.

Executed early in an engagement, a successful boarding action preserved the sails, rigging, and hull of both victor and prize. A naval commander might determine that his ship was so out-gunned by his opponent that his best chance for victory lay in boarding the enemy. Alternatively, if a commander judged

In a fifteen-minute battle on 1 June 1813, boarders from HMS *Shannon* easily secured USS *Chesapeake*'s surrender after the latter's demoralized crew succumbed to the enemy's superior gunnery. *U.S. Naval Academy Museum*

that the enemy's complement had been significantly weakened by casualties or appeared demoralized, he might lead a boarding action as the quickest way to secure victory, as in the case of HMS *Shannon*'s capture of USS *Chesapeake*.

Flexibility

Naval commanders faced numerous tactical options: To hold the weather gauge or give it up; to fight at long range or short; to fire high to disable sails and rigging, as H.M. brig *Frolic* did in its encounter with U.S. sloop of war *Wasp,* or low to kill men, dismount guns, and damage the hull, as *Wasp* did in the same engagement; to fight broadside to broadside or to maneuver for a raking opportunity; to fight with iron, or to board with steel. In every naval engagement, tactics depended on situations that were seldom static. A successful commander stayed aware of the changing situation and adapted his tactics accordingly.

MJC

Ay, tear her tattered ensign down!

Long has it waved on high,

And many an eye has danced to see

That banner in the sky;

Beneath it rung the battle shout,

And burst the cannon's roar;—

The meteor of the ocean air

Shall sweep the clouds no more

Part III: **Master Level**

War of 1812: Diplomacy

Confirming America's political and economic independence dominated the new republic's foreign relations during the first twenty-five years of its existence. Fear that Britain's maritime policies robbed the United States of its honor and relegated it to a colonial status convinced the Madison administration that war was the only alternative. Ironically, diplomatic efforts to end hostilities began within weeks of the war declaration and continued sporadically throughout the conflict. External circumstances dictated the course of negotiations. While war came reluctantly to both sides, once engaged, they both anticipated a quick resolution—the United States expected the British to come to terms quickly and the latter predicted a swift military victory. Both parties underestimated the other's resolve. A combination of economic and military circumstances, in tandem with some astute American diplomacy, brought the war to an end.

American neutral rights suffered when Europe was at war in the 1790s. A brief hiatus at the turn of the century brought a short respite, but a resumption of the Napoleonic wars in 1803 marked the beginning of a steady downward spiral in America's relations with Britain and France. On the one hand, Anglo-French hostilities created a tremendous opportunity for America's merchant fleet, which filled the vacuum produced by the British loss of the carrying trade with the French and Spanish West Indies. The British government, fearing a French invasion, adopted a series of restrictive measures to protect its commercial interests. Seeking to punish American merchants for usurping their former trade and wishing to hurt the French economically, the British invoked the Rule of 1756, which stipulated that trade closed during peacetime (U.S. trade with French and Spanish colonies, for example) could not be opened during wartime, and disavowed the reexport trade, which had conferred neutral status to French and Spanish colonial goods transshipped in American vessels to American ports and then to Europe. The Americans tried to use this so-called "broken voyage" strategy to circumvent the Rule of 1756. The British, however, closed this loophole with an Admiralty court ruling in 1805 called the *Essex* decision, which defined a continuous voyage as one in which a non-neutral ship paid an import duty to a neutral country, which then transshipped the now "neutralized" goods in neutral ships.

In an effort to restore the lucrative reexport trade and to resolve other disagreements, President Thomas Jefferson sent a distinguished Maryland lawyer, William Pinkney, to England in 1806 to assist American minister James Monroe in negotiating a treaty. The pair won some trade concessions and legalized the trade derived from the "broken voyage," but could not obtain an unconditional pledge to end the British practice of stopping American merchant vessels on the open seas and removing suspected Royal Navy deserters. Jefferson considered the British practice of pressing American seamen into Royal Navy service such a volatile topic that he refused to send the Monroe-Pinkney Treaty to the Senate for ratification because that document did not address impressment. America demanded an end to the practice; Britain refused to discuss it. The two nations stood at loggerheads in 1807.

John Quincy Adams, minister to Russia when the War of 1812 began, adroitly headed the American peace commission that negotiated the Treaty of Ghent with the British, returning the warring parties to status quo antebellum. *U.S. Naval Academy Museum*

Over the next four years the French and British adopted restrictive trade policies, and the Americans retaliated with a series of economic countermeasures in an unsuccessful attempt to protect their neutral rights. In May 1806 Britain blockaded the European coast from Brest to the Elbe, and Napoleon retaliated in November with his Berlin Decree, which declared Great Britain in a state of blockade and excluded British goods from ports held by the French. The British countered with an Order in Council (January 1807) that excluded neutral trade from ports controlled by their enemies. By the end of 1807, Napoleon issued the Milan Decree, which ordered the seizure of any vessels that stopped at British ports or permitted inspection by the Royal Navy.

In addition to having its vessels seized and its goods confiscated on the high seas, the United States was also humiliated in its own home waters in June 1807, when HMS *Leopard*, on orders to search for Royal Navy deserters, fired on USS *Chesapeake* near Hampton Roads, Virginia, killing three, wounding eighteen, and removing four seamen. Violent anti-British demonstrations erupted throughout the U.S., but Jefferson adopted an economic rather than a military response. The *Chesapeake-Leopard* affair so embarrassed America that any future diplomatic negotiations required the British to renounce the practice of impressment. While the British government disavowed the episode, it never disavowed impressment, thus making this incident a cause of the War of 1812.

In the years following the *Chesapeake* incident, the Jefferson and Madison administrations implemented a series of trade restrictions to gain the belligerents' attention because normal diplomatic channels were unproductive. The Embargo Act of 1807 was a self-imposed measure that prohibited U.S. vessels from sailing to any foreign port. The act, which failed to change British or French economic measures, hurt the nation economically and the Republican Party politically. On 1 March 1809, a lame-duck Congress, stunned with Federalist gains at the polls, replaced the Embargo Act with the Non-Intercourse Act, which restricted trade only with Great Britain, France, and their colonies, and offered to reopen trade with whomever rescinded their restrictions against the U.S.

The United States continued to negotiate its differences with Great Britain, but British intransigence resulted in lost opportunities. In April 1809 President James Madison thought he had an agreement with David M. Erskine, the British minister to Washington, to reopen trade with Great Britain, but George Canning, the British foreign secretary, repudiated his minister. Erskine obtained from the United States the substance of his instructions, but Canning wanted America formally to acknowledge its capitulation to British terms. Further negotiations faltered and in May 1810 the administration adopted yet another measure, Macon's Bill No. 2, which lifted all limits on trade and sought to create a wedge between the British and French by threatening to resume restrictions on whichever country refused to end its restrictions on the United States.

These measures showed the bankruptcy of America's diplomatic efforts. Grasping for any hint of success, the administration accepted the word of the duplicitous French who indicated that they had withdrawn the Berlin and Milan decrees when in fact they had not. Nevertheless, the administration let itself be duped by the French and threatened Britain with a resumption of restrictions, thus further exacerbating the situation with them.

By late 1811 war was inevitable. After exhausting all avenues for a peaceful resolution, Madison sent his war message to Congress on 1 June 1812. While the president briefly mentioned the British incitement of Native Americans on the western frontier as a justification for war, his central thrust was America's rights as a neutral nation. The slogan "Free Trade and Sailors' Rights" epitomized the American position. The Royal Navy's interdiction of European ports and its lurking off the coast of the United States in search of American sailors to impress into its service hurt America's commerce and stature. Ironically, just a few days before the United States declared war on 18 June, Parliament repealed the Orders in Council, not to placate the Americans but to appease British manufacturers whose export trade had suffered significantly.

Events in Europe during the War of 1812 proved just as important as land and naval battles between the two adversaries in determining the course of diplomacy. The Americans expected that Britain would come to terms quickly—anticipating that it would not wish to engage them while still engulfed in an international war against Napoleon. Therefore, a week after declaring war, Madison sent instructions to Jonathan Russell, chargé d'affaires in London, demanding Britain's repeal of the Orders in Council, which it had done already, and cessation of impressment. British Foreign Secretary Lord Castlereagh adamantly refused, however, to resolve the impressment question because the Royal Navy desperately needed manpower. Dismissive of American efforts to end the war so soon after declaring it, the British initially thought they could bring the Americans to terms with a minimal commitment of men and ships. Meanwhile, in late 1812, Emperor Alexander I of Russia offered to mediate. Castlereagh declined, but Madison agreed. By accepting the Russian offer, Madison forced the British to negotiate directly with the United States to stave off Russia's meddling in their affairs. Madison readily embraced Castlereagh's offer in November 1813.

The president's appointment of men of stature to the peace commission reflected his desire to have strong negotiators representing the American position. Conversely, Castlereagh selected undistinguished men to resolve the American war compared with the prominent men he sent to the Congress of Vienna to determine the fate of Europe. The two sides did not meet until August 1814 in Ghent, Belgium. Finally recognizing that the British would never budge on the impressment question, Madison instructed his commissioners not to press that issue. Confident that news of anticipated military victories in America would be forthcoming, the British, who sought security for their North American possessions, demanded a peace based on territory possessed at the end of the war. News of defeats at Plattsburgh and Baltimore in September 1814, however, coupled with fear that the American commissioners would end negotiations, necessitating yet another military campaign, and concerns with maintaining the balance of power in Europe after the fall of Napoleon, all left the British government more amenable to negotiate with America rather than to dictate terms.

The Treaty of Ghent, signed on Christmas Eve 1814, resolved none of the neutral rights issues that precipitated the war. These issues were moot as the conclusion of the Napoleonic wars ended Britain's need for manpower and trade restrictions. Status quo antebellum or a return to the prewar situation became the basis of the peace. While securing no territory or recognition of neutral rights, the United States did survive with its sovereignty intact and its independence recognized. The nation quickly forgot the earlier military ineptness in the euphoria of bringing Napoleon's vanquisher to a draw. The American diplomats at Ghent deserve credit for their steadfastness.

CFH

War of 1812: Politics

Although the United States declared war against the United Kingdom in 1812 in defense of freedom of the seas and sailors' rights, those Americans who were most directly interested in seagoing commerce, the New England merchants, were most passionately opposed to the war. Before 1807 Americans engaged in overseas commerce had been able to make profits despite impressment of their seamen, British Orders in Council that forced their merchantmen to touch at British ports and pay fees before heading for the European Continent, and Napoleon's decrees that ordered the seizure of any neutral ships complying with the British regulations. Jefferson's 1807 Embargo, by putting a total stop to their overseas trade, halted those profits. By depriving them of overseas markets, the Embargo hurt western farmers and southern planters, too, but despite the agricultural depression, the majority of westerners and southerners remained loyal to Jefferson's Republican Party.

Issues of relations with Indians and with countries on the nation's borders fuelled anti-British sentiment in the West and South. As the prospects of war with the United States increased, British authorities in Canada strengthened friendships with Indian tribes and provided them increased supplies. When war with a pan-Indian confederacy led by the Shawnee Tecumseh broke out in the United States in 1811, American frontiersmen blamed the British and called for conquest of Canada to end the Indian trouble. Meanwhile, southerners longed for the conquest of Florida, in the hands of Spain, a British ally, where runaway slaves took refuge and hostile Indians found safe haven, and whose rivers provided access to the Gulf of Mexico. In 1810 voters in the West and the South sent a number of warlike Republicans, known as War Hawks, to the House and the Senate.

In June 1812, convinced that the national honor required it, President James Madison requested a declaration of war. Congress complied by a narrow vote—nineteen to thirteen in the Senate and seventy-nine to forty-nine in the House. The vote followed party lines more closely than it did regional ones. Republicans were inclined to trust the president; Federalists were not. In the presidential election that winter, electors from the northeast voted for the peace candidate, De Witt Clinton, while those from the South and West voted for Madison, who retained office with an electoral vote of 128 to 89.

As the war dragged on, opposition to it gained strength. Federalists in Congress resisted war measures. Moneyed-men in the Northeast, who controlled most of the nation's cash, refused to buy government bonds for support of the war. In New England, opposition to the war led to official acts of obstruction, such as local ordinances restricting public recruiting and extending the period of quarantine for returning privateers. On several occasions the refusal of governors of New England states to allow state militia to serve outside their states scuttled military offenses into Canada. Private acts of opposition extended as far as the passing of intelligence to the enemy.

Republicans, who were responsible for the conduct of the war, entered the autumn of 1814 with a sense of foreboding. Conditions had come to a crisis, and no easy solutions offered themselves. The country was on the verge of financial collapse. The U.S. Treasury secured subscriptions for only two and a half million dollars of a six million dollar loan and in November defaulted on the national debt. Runs on banks in the West and South caused them to suspend species payments, and when banks stopped accepting each other's notes the federal government had no way to transfer funds between regions.

The British blockade caused commodity prices in the South to plummet and ruined shipping interests in New England. Trade with the enemy mushroomed by sea, through British-held Castine, Maine, and across

James Madison's presidency (1809–1817) spanned the prewar diplomatic efforts to protect American neutral rights, the war years, and the peace that ushered in an era of nationalism. *U.S. Naval Academy Museum*

the Canadian border. Residents of the island of Nantucket, Massachusetts, cut off from the mainland by the blockade and reduced to starvation, had felt obliged to declare their neutrality.

Federalists increased their seats in the House of Representatives in the fall elections, and the anti-administration faction in the Republican Party grew more vociferous. While the British were sending large reinforcements to America and launching aggressive campaigns, U.S. Army recruiting fell well below needs and desertion rates reached new heights. With America's prospects for holding off the British militarily bleak, British negotiators were demanding concessions in return for peace, including: an Indian reserve in the Old Northwest, territory in what would become Maine and Minnesota, U.S. demilitarization of the Great Lakes, and the end of U.S. fishing in Canadian waters.

In New England, sentiment in favor of secession from the United States, heard earlier at the time of the Louisiana Purchase and during the Embargo, increased. On 15 December 1814, delegates from the New England states met at Hartford, Connecticut, to discuss ways to preserve their region's influence. Moderates dominated the assembly and, instead of voting for secession, proposed seven amendments to the United States Constitution designed to protect New England from the growing influence of the South and the West. Among the proposed amendments, one would have abolished the clause counting three/fifths of the unfree population that increased the proportional representation of slave states; another required a two-thirds vote to admit new states; and another, intended to break the hold of the "Virginia Dynasty" on the presidency, prohibited a president from being succeeded by anyone from his own state.

Peace saved a United States teetering on the brink of catastrophe. News of the American victory at New Orleans and of the Treaty of Ghent ending the war arrived shortly after the Hartford Convention concluded, with the effect of discrediting the Hartford meeting. The end of the war left the anti-war party in disrepute, boosted nationalism, and subdued for a time the sectionalism that had divided the nation.

MJC

War of 1812: The Land War

The U.S. Army was unprepared to meet the challenges of fighting a war with Great Britain in June of 1812. The Army at this time mustered fewer than 7,000 effectives, or roughly one-fifth its authorized strength. The corps' senior officers, many veterans of the War for Independence, no longer possessed the energy and military élan that had characterized their service in the Revolution. Its junior officers were more distinguished for their political connections and lack of combat experience than their ability to lead men in battle. Its rank and file was poorly trained, ill-equipped, and more accustomed to duty as a frontier constabulary than to the disciplined routine of a professional fighting force. The Army's civilian head, Secretary of War William Eustis, was a weak administrator, unable to direct military affairs with force and decision. Congress's failure to provide him with the necessary staff and funds to organize the nation's war effort compounded Eustis's shortcomings as a department head.

Despite the American military's numerous handicaps, its situation was not entirely hopeless. While the British army dwarfed the U.S. Army in size, the majority of England's troops were stationed in Europe, leaving only 5,600 redcoats to defend Canada. Like the American army, these British soldiers were not particularly well officered or well trained. The 71,000 men serving in Canadian militia units (compared to the 150,000 volunteers and state militia the American government hoped to field) were also of questionable value. Governor General Sir George Prevost described the militia of

Lower Canada as "a mere posse, ill arm'd, and without discipline." Another potential source of manpower, Native Americans, though first-rate fighters, were unreliable allies, choosing to fight alongside the British only so long as it served Indian purposes. The weak and scattered state of British garrisons in Canada, coupled with their initial logistical isolation from England, presented the United States with a golden opportunity to bring the war to a swift conclusion in one campaigning season. All depended on the ability of the American army to launch a series of coordinated attacks along the Canadian frontier, overwhelming the enemy before England could bring its superior resources to bear in the conflict.

The American strategy for 1812 called for an invasion of Canada on three widely separated fronts: in the west from Detroit, in the center across the Niagara River, and in the east up Lake Champlain. The U.S. offensive began auspiciously enough with William Hull's crossing the Detroit River into Canada on 12 July and occupying the town of Sandwich. Five weeks later, in a stunning reversal of fortunes, Hull not only had retreated across the river to Detroit, but also had surrendered that post and his command to the enemy without firing a shot against his British besiegers. The loss of Detroit, along with American outposts at Dearborn and Michilimackinac, left the region west and north of Ohio in enemy hands. Fresh disasters awaited American arms on the Niagara frontier, where Brigadier General Stephen Van Rensselaer launched an attack across the Niagara River in October. Though initially successful in occupying Queenston, American forces were ultimately routed because the New York State militia refused to cross into Canada to support Army regulars. Major General Henry Dearborn led the final operation against Montreal in November. His march on that Canadian city lasted all of one week. Dearborn blamed his expedition's failure on supply shortages and the unwillingness of the militia to enter Canada. American ineptitude and superior British generalship characterized the 1812 land campaign. Never again would the U.S. military enjoy such an opportunity to land a knockout blow against its British foes.

The defeats of 1812 led to a shakeup in the War Department's leadership, with the more dynamic John Armstrong replacing William Eustis. The 1813 campaign opened with renewed American efforts to carry the fight to Canadian soil coupled with a shipbuilding program to secure naval mastery of the lakes. In April joint forces under Dearborn and Commodore Isaac Chauncey landed on the northwest shore of Lake Ontario at York (present-day Toronto), the capital of Upper Canada and site of a large enemy supply depot. The burning and looting of this town by U.S. troops would later become a justification for the British burning of Washington. The Americans followed up their attack on York with the capture of Fort George in mid-May. American military success continued on the Lake Erie frontier when Major General William H. Harrison, following Perry's victory over the British fleet on 10 September, defeated a mixed force of British and Native Americans at the Battle of the Thames on 5 October. The war on the Canadian border concluded with several disappointing setbacks including the failure of a twin thrust at Montreal up the St. Lawrence River and Lake Champlain led by Major Generals James Wilkinson and Wade Hampton in November, and the abandonment of Fort George in December.

The arrival of additional troops, sailors, marines, and ships allowed the British to pursue limited offensive operations against the U.S. in 1813. Though attacks on Sackets Harbor, New York, and on Norfolk, Virginia, failed, Rear Admiral George Cockburn led numerous

Raised a Quaker, and lacking formal military training, Jacob Brown was an unlikely candidate to lead an armed force of any size. But by war's end, the New Yorker had emerged as the nation's most successful battlefield commander. *Naval Historical Center photograph*

destructive raids in the Chesapeake Bay and along the North Carolina coast.

The last full year of fighting, 1814, proved to be the most difficult and frustrating of the war for the U.S. military. Chronic shortages of money, men, and materiel continually hamstrung operations, while command conflicts and an uncooperative spirit pervaded relations between the Army and militia. Even more alarming was the increased size of Britain's armed forces in North America. The number of redcoats in Canada alone stood at nearly 40,000 before year's end. With such large bodies of troops at their disposal, and supported by a strengthened fleet, British commanders were able to press American forces with aggressive campaigns on the Niagara and Lake Champlain frontiers, in the Chesapeake Bay, and along the Gulf Coast. Fortunately for the Army, Secretary Armstrong had elevated younger, more able officers to general's rank. In July, two of the recently promoted, Jacob Brown and Winfield Scott, acquired honor and new respect for U.S. regulars in two fiercely fought actions at Chippawa and Lundy's Lane along the Niagara frontier. Another elevated officer, Alexander Macomb, conducted the land defense of Plattsburgh, New York, that helped turn back Prevost's invading army in September. A final Armstrong appointee, Andrew Jackson, inflicted the greatest defeat of the war on the British army at New Orleans in January 1815. Despite the successes of these officers, American arms still suffered major setbacks in all theaters in 1814, the most devastating being the capture and destruction of Washington in August.

The U.S. military's record was disappointing throughout much of the war, owing in part to the numerous obstacles that undermined its performance in the field. The nation's lack of preparedness in 1812, incompetent Army administrators, and inept generalship are a few that have already been mentioned. Other contributing factors included: a flawed logistical system incapable of delivering adequate supplies of food, clothing, and equipment; failed recruiting efforts that left the Army dangerously under strength; an inadequate tax system to finance the war; and the need to conduct operations against Canada in states politically hostile to the war. The U.S. military persevered in spite of these difficulties, earning enough victories when it counted to preserve the nation's honor and drive the British to the peace table.

CEB

Constitution's Squadron and Special Service

American naval strategy in the nineteenth century defended and advanced American enterprise, and USS *Constitution*'s active duty service from the 1820s to the 1850s promoted this naval-commercial nexus.

Mediterranean Squadron, 1821–28

The Mediterranean Squadron, established in 1801 in response to continuing infringements on American commerce by the Barbary corsairs, was the first, distant-station squadron assembled by the Navy Department. The Navy disbanded the unit in 1807, but the recurrence of attacks on American shipping in the Mediterranean during the War of 1812 forced the Madison administration to resurrect it in 1815. The Navy in the quarter century after the War of 1812 wisely distributed its limited naval appropriations among foreign squadrons, using the Atlantic as a buffer to protect the homeland.

Constitution returned to American waters in May 1815 after defeating HMS *Cyane* and *Levant* in one of the last naval engagements of the War of 1812, too late to join Stephen Decatur's squadron that was about to sail for the Mediterranean. The military mission against the Algerines succeeded so quickly that *Constitution* remained at home, and within six months the ship was placed in ordinary (out of service) from 1816 to 1820. Benefiting from a naval expansionist atmosphere, the frigate received orders in April 1820 to undergo a major overhaul that led to its shipping out from Boston on 13 May 1821 for a seven-year Mediterranean deployment, interrupted by only a seven-month stateside visit.

Constitution's three American commanders during this Mediterranean stint, Jacob Jones, Thomas Macdonough, and Daniel T. Patterson, shared similar backgrounds. Entering the Navy as midshipmen during the Quasi-War with France, they served together in the ill-fated *Philadelphia* during the Barbary Wars, but had separate commands during the War of 1812. While the command style of the officers who captained *Constitution* and specific historical events differed from one year to the next, some things never changed. The everyday routine of discipline, punishment, dueling, shipboard life, accidents, and deaths was counterpoised with the official duties of showing the flag from port to port, ferrying American diplomats, and conducting diplomatic negotiations.

After an expeditious twenty-one-day transatlantic passage, *Constitution* reached Gibraltar in early June 1821 and concentrated on making the port rounds in the western half of the Mediterranean before entering

winter quarters at Port Mahon, Minorca, just before Christmas. Aiding an American merchant ship in distress, performing ship-handling drills, and receiving long-awaited mail contended with dinner parties for American consuls and immoral shore activities among crewmen in filling the daily routine. The layover at the repair depot was punctuated by a duel that left one midshipman dead and another suspended from service for six months—evidence that the cult of honor still flourished in the Navy.

On 12 March 1822 the squadron, composed of the flagship *Constitution* and its consorts (sloop of war *Ontario* and schooner *Nonsuch*) began another season of cruising. The ships headed first westward to Gibraltar and then eastward to the Aegean to ensure the safety of Americans residing in Greece in the midst of the brutal conflict raging there between the Greeks seeking independence and their Turkish overlords. While en route to Smyrna (Izmir), Turkey, in June, the Americans witnessed from afar the destruction of a Turkish ship of the line. After cruising westward to Gibraltar, then laying over at Port Mahon for two weeks of repairs, *Constitution* began a counterclockwise round of port calls, sometimes with the squadron and other times alone. All the ships returned to Minorca for the winter season of overhauls, which lasted from December 1823 to April 1824. The monotonous cruising from port to port created disciplinary situations demanding strong leadership. The log entries note the addition of a weekly practice at general quarters—perhaps the easygoing Captain Jacob Jones had begun to recognize that idleness was having a deleterious effect on his crew.

After one more season in 1824 of port calls and ferrying diplomats, *Constitution* learned in November that the Navy Department had recalled Captain Jones. The station needed a disciplinarian and a man of stature to deal with burgeoning American commercial interests in the Mediterranean and with a Turkish-Greek war that had escalated from a localized conflict to an international problem threatening the balance of power in the region. The department selected Captain John Rodgers as squadron chief, but delays in outfitting his flagship hindered him from assuming command immediately. On 9 April 1824, Captain John Orde Creighton in *Cyane* (the British sixth rate that *Constitution* captured during the War of 1812) replaced Jones as temporary squadron commodore. The next day *Constitution* sailed for New York where its new commander, Thomas Macdonough, superintended its refit before returning to the Mediterranean station in November 1824.

As part of his agenda to modernize and discipline a navy that had become lax during peacetime, Secretary of the Navy Samuel Southard appointed Macdonough to redress the permissive conditions that had arisen in the fleet during Captain Jones's command. Macdonough had honed his organizational skills during the War of 1812 by building a fleet on Lake Champlain and then defeating the British squadron at the Battle of Plattsburgh Bay in September 1814. Now *Constitution*'s new commander fashioned his Sailors into an orderly state—curtailing the frequency of liberty and punishing transgressions more regularly. The crew of the frigate spent the winter and spring of 1825 following a familiar routine—making repairs and transporting diplomats.

Captain Jacob Jones, whose USS *Wasp* bested HMS *Frolic* early in the War of 1812, commanded the Mediterranean Squadron from 1821 to 1824 with *Constitution* as his flagship. *U.S. Naval Academy Museum*

Rodgers and his prestigious flagship, the 74-gun *North Carolina*, arrived at Gibraltar on 30 May 1825 with orders to normalize relations with Turkey and to discipline the squadron. While unable to contact his Turkish counterpart during his first sailing season, Rodgers forged his squadron into a respected fighting force, one remembered for its humanitarianism in helping to combat a major fire in Smyrna.

Suffering from consumption, Macdonough asked Rodgers to relieve him from his command in October 1825. Daniel T. Patterson, who commanded U.S. naval forces at the Battle of New Orleans in January 1815, captained *Constitution*, except for a brief change of command during the winter of 1825–26, for the next thirty-three months (thirty-one months in the Mediterranean). Old Ironsides traveled from port to port in the Ottoman Empire during these years, displaying American naval might, most of the time as part of the squadron, but also as its flagship for a brief period in 1827. After a long tour of duty, *Constitution* returned to Boston in July 1828. Eventually the Mediterranean Squadron's tedious rounds of port calls resulted in the 1831 ratification of the first commercial treaty between the United States and the Ottoman Empire.

Mediterranean Squadron, 1835–38

Eight years of inactive service (in ordinary or undergoing restoration) passed before *Constitution* returned to the Mediterranean—this time regaining its flagship status under the command of another officer who served in the War of 1812, Jesse Duncan Elliott. Both Elliott and *Constitution* proved to be survivors. Elliott weathered the controversy surrounding his lack of aggressiveness at the Battle of Lake Erie while *Constitution* endured decommissioning, debates over its scrapping, arguments over its figurehead, and a hurricane and iceberg during two transits of the Atlantic.

During the first half of 1835, *Constitution* conveyed Edward Livingston to France to negotiate outstanding claims that America had against that country, returning the American minister to New York in June. By August the frigate proceeded again to the Mediterranean, arriving at Port Mahon on 19 September. Instead of wintering at that naval depot, Elliott decided, after relieving Patterson as squadron commander, to tour the eastern Mediterranean and then proceed westward to Portugal before closing the circle at Minorca in April 1836. Elliott was not one to idle long in port and he set out for another round of port calls in May, the highlights of which included a race between Old Ironsides and an Egyptian flagship, which the 1812 frigate bested easily. The nomadic squadron did not return to Port Mahon until January 1837 for a two-month stay.

Commodore Elliott's personal interests superseded diplomatic duties during *Constitution*'s final eighteen months in the Mediterranean. Elliott transported Andrew Jackson's secretary of war, Lewis Cass, to Egypt in 1837 to see if the Egyptians would accept a commercial agreement, but nothing substantive came of the mission. Collecting rather than negotiating interested Elliott in his last months on the station. Scouring the Holy Land for artifacts, Elliott stuffed every nook and cranny of *Constitution*'s hold with antiquities, including art, coins, and even a mummy. Not satisfied with these,

While at Malta in February 1838, *Constitution*, the Mediterranean Squadron's flagship, celebrated George Washington's birthday flying flags of all nations from its rigging. *U.S. Naval Academy Museum*

Elliott decided to convert part of the frigate's gun deck into stalls to house the horses, jackasses, and hogs that he rounded up at Port Mahon. In this non-naval atmosphere, it is no wonder that discipline broke down on board *Constitution* and that a near mutiny occurred on the ship's return to Virginia on 31 July 1838. Misconduct charges issued by two subordinate officers led to Elliott's suspension from the service for four years.

Pacific Squadron, 1839–41

After spending about ten years in different deployments in the Mediterranean during the 1820s and 1830s, *Constitution* rested in ordinary from August 1838 to March 1839, before being designated Commodore Alexander Claxton's flagship on the Pacific Squadron station. While the weather, languages, geography, and political issues would differ from those Old Ironsides was familiar with in the Middle Sea, the ship's paramount duty remained the same: protection of American commerce. Now its main cruising grounds encompassed the west coast of South America, interspersed with occasional visits to the California coast and the Sandwich (Hawaii) and Society (Tahiti) Islands, but the ship never ventured beyond the west coast of South America. The Pacific Squadron at this time varied in strength from three to five vessels.

Constitution was on station at Valparaíso on 2 November 1839 after leaving New York on 20 April. The State Department frequently took advantage of naval vessels proceeding to their stations to ferry its ministers to foreign posts. *Constitution* deposited the minister to Mexico, Powhatan Ellis, at Veracruz before proceeding to its new station. This detour, as well as stops in Havana and Rio de Janeiro, relieved the boredom of months at sea. Inactivity bred problems among officers as well as the enlisted. Captain Daniel Turner, commander of *Constitution*, clashed with Commodore Claxton during the voyage over the latter's minor interference with shipboard affairs. Claxton's death from dysentery in March 1841 resolved their differences.

Activities such as reading and staging theatrical productions helped to wile away the tedium of shipboard routine. The ship's library contained several hundred volumes purchased by the crew while in New York, and many popular authors of the day were represented. While at Rio de Janeiro during the outbound passage, the Constitutions bought theatrical costumes, sets, and props, and later they performed several plays, some that the men wrote.

Drunken fistfights were the closest the crew came to combat, although the threat of a war with Great Britain placed the frigate on high alert while it was anchored at Callao, Peru, in July 1841. Captain Turner received an official report that the British government threatened to issue a war ultimatum if a Canadian accused of murder was not released. But the alarm was short-lived and *Constitution* proceeded homeward on 11 July 1841, stopping in Rio to restock, engage in naval salutes, and embark about twenty-five Sailors from the Brazil Squadron. By the end of October *Constitution* had anchored in Hampton Roads and proceeded to the Gossport Navy Yard for four months in dry dock.

Home Squadron, 1842–43

Tension between the United States and Great Britain over the African slave trade, the status of Texas, and boundary disputes in the Northeast and Northwest, all brought the two countries close to war in the winter and spring of 1840–41. The post-War of 1812 Navy, as a cost-saving measure, depended on the oceans to protect the American homeland and used its naval appropriations to maintain squadrons dispersed around the world. The crisis with Britain, however, generated concern that America was defenseless against a domestic attack. Secretary of the Navy Abel P. Upshur recognized that the mobility and shallow draft of foreign steam warships had rendered the country's shoreline more vulnerable than before to seaborne assault. Thus the Navy established the Home Squadron in 1841 with vessels ranging in size from the flagship *Independence*, a 54-gun razee (a vessel reduced by a deck), to the 10-gun steamers *Missouri* and *Mississippi*. Dispersing the ships among several ports gave greater geographic coverage in case of an attack. The Navy Department assigned *Constitution* to Norfolk. No sense of urgency drove the preparations to ready the frigate for sea, as a full year of repairs, outfitting, and manning transpired before its commander, Captain Foxall Alexander Parker Sr. ordered "anchors aweigh." Water leaks in the officers' staterooms cut the cruise to a short three weeks and ended *Constitution*'s very brief service in the Home Squadron. The problems discovered when the ship returned to Norfolk on 2 December 1842 were so substantial that the Navy Department postponed the multiyear special cruise it had planned for Old Ironsides.

World Cruise, 1844–46

The naval policy of President John Tyler's administration was two-pronged—defense of the coastline, which the establishment of the Home Squadron addressed, and promotion and protection of commerce, to which all the other squadrons attended. In an age of western expansionism at home and a growing interest in America's manifest destiny abroad, a sixty-five-year-old naval captain in need of a job and a forty-seven-year-old ship in need of a second chance joined forces to promote America's national interest.

The Navy Department charged Captain John Percival with protecting and expanding American overseas trade during his twenty-seven-month world cruise in *Constitution* from 1844 to 1846. *Naval Historical Center photograph*

In the early 1840s, Captain John Percival (*Constitution*'s oldest commander ever), who had spent his entire career in either the merchant service or the Navy, desperately wanted a ship to command in order to support his family after losing his life's savings in a failed bank. The best the Navy could offer Percival was overseeing the conversion of ship of the line *Franklin* to a receiving ship. Meanwhile, the Navy Department planned a special, multiyear service for *Constitution*, but estimates to ready it for this arduous duty were $70,000, far exceeding the Navy's resources. On his own initiative, Percival visited the ship and vowed that he could overhaul it for $10,000. Much to the chagrin of his detractors, Percival succeeded and commissioned his ship in March 1844. However, waiting for the arrival of Henry A. Wise, the newly appointed minister to Brazil, delayed the commencement of the cruise until late May 1844. Percival's orders reflected the administration's views that maritime commerce was the lifeblood of the nation and it was the Navy's duty to promote foreign trade in peacetime and defend it in wartime. Specifically, this translated into aiding commercial enterprise by charting new anchorage sites and protecting American shipping in foreign countries. Additionally, Percival was charged with scouting out future naval coaling depots and with showing the Stars and Stripes as a counterweight to the British Union Jack.

The twenty-seven-month cruise began auspiciously with fair winds and calm seas as *Constitution* set its course on 29 May for Rio de Janeiro. After stops in the Azores, Madeira, and the Canary Islands, where the local consulates held parties and the crew of the frigate reciprocated with festive dinners, *Constitution* approached the equator on 23 July. At the appointed hour, King Neptune boarded the ship to welcome all those "polliwogs" who had never crossed the line. Minister Wise's bribe of a keg of spirits spared his family from the revelers' initiation rites.

After getting Wise and his entourage of thirteen family members and servants safely to Rio in early August, Captain Percival remained a month there replenishing his food stocks, knowing full well the importance of a healthy, contented crew to the success of a cruise. Without discipline, life on board a man-of-war could degenerate quickly into chaos, so the captain demonstrated his authority on 8 September as the ship cleared Brazilian waters by punishing nine crewmen for infractions incurred while on liberty at Rio. Weathering a gale in the South Atlantic, *Constitution* rounded the Cape of Good Hope and proceeded on a northerly track, hugging the eastern coast of Africa. The frigate stopped at Madagascar, Mozambique, and Zanzibar to restock its water and wood. After surveying the coastal commercial activity of American merchants, Percival recommended that the Navy Department station shallow-draft naval vessels in the Indian Ocean to protect its interests.

The five-week journey of 4,500 miles across the torrid Indian Ocean taxed the stamina of all the Constitutions. Percival, himself severely incapacitated with gout and other maladies, relinquished his day cabin to accommodate the swell of patients suffering from dysentery. While the medical situation on board *Constitution* in December 1844 was serious, all agreed that the potential effects of spending 120 days in the tropics could have been worse but for Percival's clever idea while in Rio to repaint the frigate's black hull a light lead color, thus deflecting some of the scorching heat ranging in temperature from 80 to 100 degrees.

Constitution found itself at the beginning of 1845 in the Far East. By the end of the year it was off Monterey Bay, California. In between, the ship experienced both the dull and the tempestuous sides of shipboard life. From January to May 1845, *Constitution* visited Sumatra, Singapore, and Borneo, where its crew found that adverse winds and currents, coupled with inaccurate charts and the frigate's deep draft, resulted in some harrowing passages while island-hopping. A page from *Constitution*'s illustrious fighting days during the War of 1812 was turned when Captain Henry Ducie Chads,

RN, visited the ship while it was anchored in Singapore in February. Chads recalled that he had surrendered H.M. frigate *Java* to William Bainbridge, *Constitution*'s commander during that December 1812 engagement. The American crew practiced small arms drills and amphibious assaults and withdrawals because of the well-known piratical threat in Borneo. Boat expeditions up the Sambas River failed to expand commercial opportunities for the United States. After dodging reefs and shoals for days, *Constitution* turned north, leaving Borneo astern.

Perhaps because no important treaties were negotiated during this around-the-world cruise, the events that occurred during the sixteen days *Constitution* anchored at Tourane (Da Nang), Cochin China (Vietnam), defined the voyage. A humanitarian effort soured when Captain Percival failed to recognize the reality of power politics in a foreign country. On 14 May a group of "Chinese" (what the Vietnamese were called then) toured the frigate and one of the visitors left behind a letter from a French missionary, Bishop Dominique Lefevre, who, along with twelve other Vietnamese converts, was imprisoned and sentenced to death. With eighty, well-armed Sailors and Marines from his crew to support him, Percival immediately demanded that he be allowed to correspond with the Frenchman. Percival set a twenty-four-hour deadline, took three local leaders hostage, and threatened to destroy the forts and shipping in the harbor if his demands were not met. The local authorities coyly ignored the Americans, buying themselves time to bolster their military position. Finally, Percival realized that *Constitution* could not defend itself, let alone overpower the aggregate power being marshaled against it. Old Ironsides, rather ignominiously, sailed after sunset on 26 May for China.

Constitution summered in several Chinese ports, but Percival missed the chance to negotiate a "most favored nation" treaty, which came eventually with the 1844 Treaty of Wanghia. Consequently, Percival's Chinese mission consisted of listening to complaints from American merchants about their commercial problems and watching the sick list soar from a recurrence of dysentery among the crew.

Constitution spent most of September 1845 in the Spanish Philippines where the most noteworthy occurrence was provisioning a British squadron, which was desperately in need of supplies. The homesick American crew considered that its entering the Pacific on 28 September presaged the beginning of the trek back to the United States. Having weathered two gales before reaching Hawaii in mid-November, *Constitution*'s company was disappointed to learn that a war with Mexico was brewing and that their ship had been ordered to join the Pacific Squadron. Leaving Honolulu on 2 December, *Constitution* met that squadron on 13 January 1846, but Percival convinced Commodore John D. Sloat that his vessel desperately required repairs, and he received permission on 22 April to return Old Ironsides to the states just days before the Mexican War broke out.

Leaving California, *Constitution* headed south, stopping at Valparaíso for supplies before rounding Cape Horn on the 4th of July en route to Rio de Janeiro. There, Percival learned that America and Mexico were at war. *Constitution* convoyed a number of merchant vessels, eager for the big frigate's protection, safely to the Delaware Capes, and then it returned to Boston on 27 September 1846, having sailed 52,370.5 miles.

Constitution's victories during the War of 1812 marked the independence of the young republic. By the 1840s its worldwide cruise signified America's expansive quest for economic independence by promoting American commercial interests overseas.

Mediterranean Squadron, 1848–51

In 1848 Revolutionaries in France, Germany, the Austro-Hungarian Empire, and Italy struggled to establish constitutional rule and representative assemblies throughout Europe. Responding to the turmoil caused by these almost simultaneous revolutions, the Navy increased its naval presence in the Mediterranean by sending *Constitution* for a third and final mission to that region. After outfitting and manning *Constitution*, Captain John Gwinn, a War of 1812 veteran, got the frigate under way on 9 December 1848. Arriving at Tripoli on 19 January, Old Ironsides took aboard Consul Daniel Smith McCauley and his family and transported them to Alexandria, Egypt. Mrs. McCauley delivered a son while en route, naming him Constitution Stewart in honor of his great uncle, Charles Stewart, *Constitution*'s commander at the end of the War of 1812.

A similar routine of making port calls, ferrying diplomats, repairing the ship, and assisting Americans in distress filled the frigate's days. Not everything remained the same, however. La Spezia, Italy, had recently replaced Port Mahon as the Mediterranean Squadron's repair depot because the antics of the American tars while on liberty prompted the Spanish to cancel the leasing agreement at Minorca.

Captain Gwinn's crew considered him a martinet for his record number of floggings, and the squadron commander, Commodore Charles W. Morgan, deemed him insubordinate for disobeying orders to remain strictly neutral in disputes between the Italian insurgents and Ferdinand II, King of the Two Sicilies, and Pope Pius IX. On 4 September 1849, the unpopular captain died of chronic gastritis and was buried at Palermo, Sicily, while many of the ship's company drank heartily to his demise. During 1850, *Constitution* continued routine squadron duty under its new commander, Captain Thomas S. Conover, making the rounds from Genoa, Toulon, Leghorn, Marseilles, Naples, and La Spezia, before beginning its voyage home on 2 November. The Atlantic passage was uneventful, except for *Constitution*'s collision one night with *Confidence*, a British brig that sank instantly, presumably with all hands. After a fruitless search for survivors, the frigate got under way only to discover that all but one of the ill-fated ship's crew had managed to attach themselves to the American ship and scramble aboard to safety. Returning to the New York Navy Yard on 11 January 1851 after a two-year deployment, *Constitution* found its future a subject of discussion within the Navy. To save on costly repairs some advocated striking it from the Navy list and converting it into a memorial ship. Once again Old Ironsides would dodge the ship-breaker's yard.

Africa Squadron, 1853–55

Although slavery would exist in the United States until the end of the Civil War, Congress had banned the importation of slaves as of 1 January 1808. Subsequent congressional legislation in 1819 authorized the president to use naval vessels to seize American ships engaged in the African slave trade. Beginning in 1820, the Navy Department sent ships to the African coast occasionally, but not until 1843 was the American naval presence there formalized with the Africa Squadron. In the 1842 Webster-Ashburton Treaty, Great Britain, the leading opponent of the slave trade since 1807, convinced the United States that each country should maintain separate squadrons of ships off western Africa with ordnance totaling at least eighty guns.

During most of the pre-Civil War years, sloops of war served as flagships of the Africa Squadron. While designating the frigate *Constitution* as the squadron's flagship in 1853 saved the ship from becoming a floating memorial in Boston, that action also indicated the ship's reduced status. In addition, for the first time a commander, John Rudd, would be in charge of Old Ironsides, not a captain. Duty on the African station was the least desirable of all the commands, but operational service was preferable to "swinging on a hook" for the fifty-five-year-old ship.

Two sloops of war and a brig also comprised the Africa Squadron but they served independently of *Constitution*, the flagship of Commodore Isaac Mayo. When the frigate set out on 2 March 1853 on its last operational cruise, it could expect to perform the same duties that it had undertaken on other stations: ferrying diplomats, protecting American commercial interests, negotiating agreements, stopping for repairs, and socializing at some ports. The main purpose for establishing this station, however, was to stanch the flood of vessels carrying Africans to slavery in the Americas. While cruising near the Angolan coast, *Constitution* chased but did not catch a slaver on 26 October 1853. A week later the frigate succeeded in bringing to the American schooner *H. N. Gambrill* that had obviously been used for transporting slaves across the Atlantic. This slaver would be the last prize *Constitution* would ever take.

While *Constitution* made only this one capture, it attained several diplomatic achievements during its twenty-seven-month cruise. Asked to intercede between two warring tribes in Liberia, Commodore Mayo sent a contingent of men and ordnance ashore and shelled a town, which brought all parties together to negotiate in the frigate's poop cabin. Mayo's success in mediating the Barbo-Grebo Peace Treaty on 6 September 1853 led to another diplomatic feat in July 1854 when the commodore negotiated a territorial dispute between the Grahway and Half Cavally tribes.

Constitution had its share of lighter moments when its crew could cast aside the often arduous and boring routine of shipboard life. For instance, there was time for sightseeing at Saint Helena (the exiled Napoleon's last residence after his defeat at Waterloo), parties ashore, and galas and theatrical productions on board the frigate when it anchored at Madeira or at its base of operations, Porto Praia in the Cape Verde Islands. Fireworks highlighted July 4th celebrations and on 19 August 1854 Old Ironsides commemorated the forty-second anniversary of the frigate's defeat of *Guerriere* by "splicing the main brace."

By the end of March 1855, *Constitution* ended its tour with the Africa Squadron and weighed anchor for home. While in transit, Commodore Mayo, learning of an alleged incident between a Spanish warship and a U.S. mail steamer off Cuba, diverted *Constitution* to that island to assist if necessary. There was no crisis and the American warship returned to Portsmouth, New Hampshire, on 2 June 1855—its operational days over.

CFH

Service as a Training Ship

In 1794 Congress authorized the construction or purchase of six frigates. Sixty years later in 1854 it approved the building of six steam frigates with screw propellers. While naval sailing ships would still grace the world's oceans, their end was in sight, as the U.S. Navy by the 1850s had committed itself, after twenty years of experimentation, to steam. When *Constitution*

From 1857 to 1871, *Constitution* served as a stationary school ship for midshipmen, giving young naval officers in training their first taste of shipboard life. Except for a brief wartime relocation to Newport, Rhode Island, the historic frigate acted as a floating classroom while tied to an Annapolis wharf. *Naval Historical Center photograph*

returned from its two-year African cruise in June 1855, it had seen almost fifty-eight years of naval service as a sailing warship. Miraculously, it had survived an attempt in 1845 to convert it to a paddle vessel. Now it assumed an educational role and endured as the iconic symbol of an era of naval glory.

Historians of the mid-nineteenth-century Navy have found that midshipmen who learned their craft before the establishment of the Naval Academy were better schooled in shipboard training, while those "young gentlemen" who attended Annapolis studied more seamanship and navigation in a classroom than on a ship. Indeed, for its first six years the Academy lacked a summer practice school or drill ship for teaching seamanship or great gun drill. A Navy Department regulation remedied this discrepancy in 1851 by requiring that the school maintain "a suitable vessel of war" for sea service and gunnery practice and embark the midshipmen on a summer cruise each year after their June examinations. During the 1850s, midshipmen cruised in sloops of war *Preble* and *Plymouth*.

After returning from its Africa Squadron cruise in 1855, *Constitution* remained in ordinary on the ways at Portsmouth Navy Yard for two years. Meanwhile, in 1857 the Navy Department decided that the shore-bound midshipmen studying at the twelve-year-old Naval Academy at Annapolis, Maryland, required more classroom space and hands-on training in shipboard life. The Navy ordered Old Ironsides to serve as a permanent school and drill ship for the midshipmen. Before assuming this new task, *Constitution* underwent a three-year structural overhaul that included several additions—recitation rooms in the poop cabin and spar deck, study rooms running the length of the gun deck, lockers on the berth deck, and wash rooms forward. Recommissioned as a second-rate ship on 1 August 1860, *Constitution* now carried only sixteen guns.

After leaving Portsmouth on 5 August, the big frigate negotiated the narrower and shallower confines of the Chesapeake Bay and Severn River, finally being towed to an anchorage off the Academy on the 20th. All 127 new fourth classmen lived, studied, and worked in the ship, away from the supposedly bad influence of the upperclassmen.

While *Constitution*'s first class of midshipmen embarked on its first year of naval training, the nation experienced the unsettling ramifications of Abraham Lincoln's election and the resulting disaffection of the Southern states with the Union. Maryland harbored many southern sympathizers and its possible secession threatened to jeopardize the safety of the Academy and *Constitution*. Rumors abounded. By April 1861, Secretary of the Navy Gideon Welles issued orders to defend Old Ironsides or destroy it. The Naval Academy's commandant directed its crew of students, Sailors, and Marines to prepare the ship for an assault. Vulnerable from land and water, *Constitution*'s men established watches and repositioned the ship's 32-pounders. In the early morning darkness of 22 April, the vigilant midshipmen almost attacked an unsuspecting ferry that was carrying militia troops to Washington until disclosure of the vessel's Union identity at the last moment spared it. As this incident accentuated the frigate's precarious position while moored at an Annapolis wharf, the Navy decided to enlist the aid of the ferry and its soldiers in hauling *Constitution* out to a deep-water anchorage where escape would be easier. In the span of twenty-four hours, the frigate survived a false alarm of enemy ships nearby and overcame grounding on a mud bank and a bar by kedging free.

Unable to locate a safe, shore establishment for the midshipmen and a suitable anchorage for *Constitution* in the Chesapeake region, the Navy Department decided to send the school and its training ship to Newport, Rhode

Island, while the Civil War raged. Under a tow and escort, *Constitution* carried 140 midshipmen, their baggage, and scholastic material northward. Soon after arriving at Newport on 9 May, the midshipmen experienced the harsh reality of war when the Navy began calling upperclassmen to active duty. The government-leased space in Newport, Atlantic House, served both as the Academy's headquarters and residence for upperclassmen, while the fourth classmen continued to live in *Constitution*, moored at Goat Island.

This only known photograph of *Constitution* under sail while in regular service was taken in 1881 during Old Ironsides' last year as an apprentice training ship. *Naval Historical Center photograph*

Tied to a wharf, off Newport, *Constitution* served as a school rather than a cruising ship, until the end of the Civil War when the entire Academy establishment returned safely to Annapolis. On 9 August 1865 the sixty-eight-year-old *Constitution* stood out from Rhode Island with its sails set but only to assist its tug, the steamer *Mercury*. Old Ironsides was soon cut free on the Atlantic leg of the trip because it was able to outdistance its tow by averaging seven to nine knots. *Mercury* resumed its towing duties for the transit from Hampton Roads, Virginia, to Annapolis. *Constitution*'s sailing days were numbered as the overhaul that it endured on its return to Maryland included connections for steam heat and gaslights—all foretelling its permanent shore status.

Routine activities filled *Constitution*'s last six years as a school ship at Annapolis. The midshipmen returned to the ship for classes in September and left for cruises in June. The frigate's crew, usually a lieutenant commander, two or three warrant officers, a paymaster, three or four mates, and thirty to sixty enlisted men, maintained the frigate and all the other training vessels. Inevitably, wooden ships require overhauls and the Navy Department decided in the summer of 1871 to end *Constitution*'s first and longest service as a training ship. The crew removed all vestiges of its educational role (the recitation house and study room) and detached most of the upper yards and rigging, leaving only a few sails needed for the towing trip from Annapolis to the Philadelphia Navy Yard. Meanwhile, the Navy debated the frigate's fate.

Five years later, after undergoing another restoration, *Constitution* was recommissioned on 13 January 1877 and the ship began its second stint as a training vessel in Philadelphia. A manpower shortage in the Navy precipitated congressional legislation in 1874 to encourage the establishment of public marine schools. Sixteen to eighteen-year-old apprentices enlisted and served in the Navy on board training vessels, receiving an elementary education and seamanship training in return. At twenty-one these Sailors transferred to cruising squadrons. For most of 1877, teenagers on *Constitution*, one of the ships selected, learned practical navigation and engaged in gun, small arms, and cutlass drills.

The Navy interrupted *Constitution*'s educational responsibilities in 1878 with a sojourn in France. The ship's crew packed, offloaded, and repacked exhibits sent to the 1878 Paris Exposition. After this year abroad (*Constitution*'s last cruise to Europe), the ship returned to Philadelphia in June 1879, shedding its freighter duties and resuming its training status.

The Navy of the 1880s stood at a crossroads. Almost twenty years of congressional neglect left much of the U.S. fleet decaying and unserviceable. *Constitution* survived this eclipse because it symbolized America's determination to defend its liberty. How fitting that its last assignment as an active unit of the Navy was to train young Sailors. For two and a half years from July 1879 to December 1881, *Constitution* served with the Apprentice Training Squadron, a newly restructured training unit that provided sea training for fifteen- to eighteen-year-old enlistees. The frigate sailed with its first group of youngsters on 19 July 1879 to New York, where the ship underwent a mandatory overhaul before setting out on a cruise on 8 October. Reaching the Caribbean by Christmas, the ship made quick island hops before turning northward on 1 March 1880. *Constitution*'s crew kept the new recruits busy with small boat exercises, and fractious boys learned about naval punishments.

Returning to New York in May, the eighty-two-year-old frigate passed a structural survey, permitting it to embark a new set of recruits on 8 July for a four-month cruise that ranged from Halifax, Nova Scotia, to Yorktown, Virginia. One of the highlights of this cruise for the boys was firing rounds from the 32-pounders. After *Constitution* returned in November to Philadelphia, the yard workers repaired its rudder, bilge pumps, and heating plants.

The Navy Department decided to expand the Apprentice Training Squadron into a fleet, homeported at a newly established naval training station at Newport, Rhode Island. *Constitution* set sail on 9 April 1881 for its new assignment but didn't reach its destination until June. The venerable frigate was detoured to the Potomac River, which it unsuccessfully attempted to ascend, to take part in funeral ceremonies in Washington, D.C., for Admiral David G. Farragut. Emergency repairs at the Norfolk Navy Yard to add iron braces to the ship delayed it even further. After reaching Newport in June, the ship participated in short training cruises over the next few months, fanning out from Newport for an American Revolution centennial celebration in New Haven, Connecticut, and anchoring off other New England ports—all the while the young tars soaked up naval jargon and ways.

The years of neglect and bad overhauls caught up with the ship, and a structural survey in November declared it unsafe for sea duty. The Navy Department of 1881 was still reeling from charges of mismanagement and lacked money to repair wooden ships, as all efforts were directed to replacing those obsolete vessels with a new steel navy. The ship's decommissioning was quickly effected. *Constitution*'s captain, Commander Edwin Shepard, transferred the apprentices to another ship in the fleet, towed it from Newport to the New York Navy Yard, and removed the rigging, stores and guns—all within a month of receiving the survey. The ship's mission to train young recruits ended with its decommissioning on 14 December 1881. Old Ironsides next served as a receiving ship in Portsmouth, New Hampshire, for Sailors awaiting assignment.

CFH

Still Afloat: Overhauls of Old Ironsides

The materials the Navy used to build its ships in the early nineteenth century—wood, hemp, canvas, metal—were all susceptible to decay. Time, weather, accident, combat, and sometimes neglect combined to damage these components, eroding a ship's seaworthiness. A ship with significant deterioration, even a recently built one, was often a candidate for being broken up or sold out of service. One of the aspects that marks *Constitution* as a singular vessel is that it has remained afloat and part of the Navy's fleet for over two centuries. That *Constitution* has endured while its sister frigates have long since passed into oblivion, and successive generations of more technologically sophisticated and powerful ships have been scrapped, is a testament to its superior design and construction, a continuing public interest in its fate, and much good fortune. Of course, *Constitution* could not have remained afloat long without the benefit of timely maintenance and repairs. The text that follows provides a brief summary of the more significant overhauls *Constitution* has experienced to date.

In the fall of 1828, the Board of Navy Commissioners ordered the naval constructor at Charlestown Navy Yard, Josiah Barker, to conduct a general survey of *Constitution*. Having recently returned from a seven-year tour in the Mediterranean Squadron, the venerable 44-gun frigate stood in need of repairs. Though Barker reported the keel and frames of *Constitution* to be in sound condition, just about every other part of the ship and its outfit needed repair or replacement. Barker estimated the cost of this refurbishment to be $113,000. For two years *Constitution* remained idle at Charlestown, its condition worsening and the cost of its projected overhaul rising by another $45,000. In mid-September 1830, with the Navy Department on the verge of issuing orders for *Constitution*'s overhaul, the *Boston Daily Advertiser* erroneously reported that the historic frigate was to be broken up. The news prompted a 21-year-old Harvard graduate, Oliver Wendell Holmes, to make a public appeal to save the ship in a poem entitled "Old Ironsides." (See poem below.) Holmes's poem, reprinted in newspapers throughout the country, generated considerable national interest in the condition of *Constitution*. Public concerns eased when the secretary of the navy issued orders on 22 September directing the frigate's repairs to get under way. But because the Charlestown Yard's new dry dock was still under construction, work on the frigate was delayed for thirty-three months. On 24 June 1833, Old Ironsides, under the honorary command of Isaac Hull, entered dry dock. Numerous state and federal dignitaries including Vice President Martin Van Buren attended the event. The

Constitution under repair at the Philadelphia Navy Yard, 1873. Navy officials hoped to have the frigate restored in time to participate in the country's centennial celebration—a hope that went unrealized. *Naval Historical Center photograph*

overhaul of *Constitution* now began in earnest under the supervision of yard commandant Jesse D. Elliott. Repairs made to the ship over the next twelve months included: replacement of the ship's interior (ceiling), exterior (strakes), and deck planking; installation of new orlop- and berth-deck beams; recaulking throughout; and a newly coppered hull. A controversial addition to the ship at this time was a carved figurehead of Andrew Jackson. *Constitution* was refloated on 21 June 1834. Work on the ship halted until mid-December when the Navy Department ordered the frigate prepared for sea. The yard's workmen had the ship completely rigged and outfitted in less than two months so that on 9 February 1835, Elliott could report *Constitution* ready to receive its crew. The first major overhaul of Old Ironsides was complete.

Nearly four decades would elapse before *Constitution* underwent its next significant renovation. The origins of this repair date to the summer of 1871, when the Navy Department issued orders placing *Constitution* in ordinary, after having logged its thirteenth year as a Naval Academy training vessel. In September the Navy tug *Pinta* towed the aging frigate from Annapolis to the Philadelphia Navy Yard where it was promptly decommissioned. Department officials took nearly a year and a half before determining the fate of the 1812 warship. Some thought was given to scrapping the frigate or even converting it to steam propulsion. But in the spring of 1873, Navy executives decided to restore *Constitution* to its original appearance and open it to the public as part of the nation's centennial celebration. That summer, project supervisors researched and drafted plans for the frigate's restoration. The following January, the ship was moved onto a sectional dock preparatory to being hauled ashore. Once the frigate was ashore in March, workers stripped *Constitution* down to its live-oak frames (futtocks). Repairs proceeded slowly because much of the equipment and personnel needed for the overhaul was being transferred to the navy yard's new site on League Island. After *Constitution* was refloated in January 1876, the Navy Department hired a private shipyard to complete the frigate's final restoration. When the contractor failed to complete the work by year's end, Navy officials took back *Constitution* unfinished. The time of the centennial's celebration having passed, Old Ironsides was placed back in commission to serve as a training ship for apprentice boys. Among the more significant changes to the ship's visual appearance during this overhaul were the substitution of a scroll-patterned billethead for the once controversial Jackson figurehead and the adoption of a simplified eagle and star design to decorate the stern.

Constitution's assignment in 1882 to Portsmouth, New Hampshire, as a stationary receiving ship ushered in a period of steady decline in its condition. Repair work performed on the frigate in 1906–7, following its transfer to Boston as a display ship, failed to halt the vessel's worsening state. A general survey of Old Ironsides conducted by the Navy Board of Inspection and Survey in February 1924 revealed just how far the ship had deteriorated since its last overhaul. According to the inspectors, *Constitution* was taking on so much water that it had to be pumped daily. Its hull was seriously distorted (the port side bulged out a foot wider than the starboard side) and its keel had a fourteen-and-a-half inch hog (bend). In addition, the ship's deck beams were decayed and many of its knees and breast hooks contained rotten wood. The stern was in such poor shape that it threatened to fall off, while defective planking had been patched with cement. Despite *Constitution*'s sad state, the Board recommended that it be rebuilt, refitted, and preserved. They estimated the necessary renovations to cost $400,000. Secretary of the Navy Curtis D. Wilbur acted swiftly on the Board's report, requesting authority from Congress to repair *Constitution*. He also asked that the restoration be paid for with private contributions rather than public funds. Congress authorized the restoration on 3 March 1925, and a national executive committee was formed to oversee the fundraising effort. America's schoolchildren responded to the call for contributions by donating $135,000 in pennies to the fund. Marines and Sailors added another $31,000, and an additional $165,000 was raised through the sale of more than one million lithographs of a specially commissioned painting of *Constitution* by Gordon Grant. Ultimately, the public donated more than $942,000 to help renovate Old Ironsides.

John T. Lord, a lieutenant in the Navy Construction Corps with knowledge of wooden ship building, supervised the overhaul of *Constitution*. He faced several challenges in repairing the ship. One was finding sufficient contemporary plans to guide the restoration of *Constitution*. Because of significant gaps in the ship's early documentary record, it was decided to restore the frigate to its 1850s appearance. A more difficult task was locating the large stocks of wood required for the intended repairs: live oak and white oak for heavy structural pieces, white pine for masts, and long leaf yellow pine for deck planking. A nation-wide search for the timber

Constitution in dry dock at the Charlestown Navy Yard, 1929. Pennies contributed by children from across America helped fund this major renovation of the ship. *Naval Historical Center photograph*

Part III: Master Level

turned up live oak from Pensacola, Florida, white oak from Ohio, West Virginia, and Delaware, and Douglas fir from Washington State, which was substituted for both types of pinewood. By the spring of 1927 enough wood and moneys had been collected for Secretary Wilbur to authorize repairs to commence. On 16 June, *Constitution* entered dry dock. A specially designed bracing system helped stabilize and support the ship's fragile hull while out of water. When the frigate emerged from dry dock in March 1930, it boasted a new keelson and assistant keelsons, a rebuilt stern, a new cutwater and bowsprit, fresh planking inside and out, and a recoppered hull. Once refloated, the ship received its masts and a newly-cast battery of replica 1812 cannon. Special internal fittings (water lines, electrical lighting, and modern toilets and stove) to accommodate the crew during the upcoming cruise were also installed. Yards, rigging, and a full set of sails were added following a ceremonial tow around Boston Harbor on 8 October. On 1 July 1931 *Constitution* was recommissioned after undergoing its most complete and extensive overhaul ever. The final cost of the restoration was $987,000.

The day following its recommissioning, Old Ironsides embarked on a three-coast tour of the United States. The frigate was commanded by Commander Louis J. Gulliver and served by a crew of six officers, sixty Sailors, and fifteen Marines. Navy Department officials hoped the tour would inspire pride in America's naval heritage and give the thousands of Americans who had contributed to the ship's renovation an opportunity to visit it. The minesweeper *Grebe* accompanied *Constitution* on its tour, towing the historic warship from port to port and providing it with electrical power when moored in harbor. Between 3 July 1931 and 16 April 1932, *Constitution* called at forty-four port cities along the Atlantic and Gulf Coasts, receiving more than two million visitors. The frigate entertained its largest crowds at New York (102,307), Philadelphia (154,809), Mobile (119,722), New Orleans (193,881), and Houston (110,406). After being refurbished and repaired at the Washington Navy Yard, *Constitution* set out on 8 December 1932 on the Pacific leg of its tour. The ship transited the Panama Canal on 27 December, arriving at San Diego on 21 January 1933. At each stop of its Pacific Coast tour *Constitution* attracted huge crowds: 180,000 at San Diego; 470,000 at San Pedro; 375,000 at San Francisco; and 500,000 in the Puget Sound area. Old Ironsides remained on the Pacific coast until 20 March 1934 when it departed San Diego for home. The ship returned to Boston on 7 May after receiving 4,614,762 visitors and traveling 22,000 miles.

The most significant overhaul of Old Ironsides' career was also its most recent, and began when the ship entered dry dock on 25 September 1992. On inspection it was found that the vessel's keel had a hog of nearly fourteen inches. The ship's restoration team, the Naval Historical Center Detachment, found the solution to this problem in the papers of *Constitution*'s designer, Joshua Humphreys. The Detachment determined that a number of Humphrey's original design elements, which had helped the ship resist hogging, had been removed during previous renovations. Restoring them would help stiffen Old Ironsides' hull, giving the frigate greater longitudinal strength and integrity, which in turn would minimize its tendency to hog. When *Constitution* emerged from dry dock on 26 September 1995, these newly installed design elements—diagonal riders, thicker, scarf-locked deck planking, additional standard knees, mid-ship knees, and stanchions—had reduced the frigate's hog to less than two inches. Other noteworthy changes made to *Constitution* during its dry-docking included: the substitution of removable for permanent cabin partitions, the addition of live oak pin and fife rails, and a reconfigured orlop deck containing an aft magazine and gunpowder filling room. Repair work on the gallant warship officially ended on 27 March 1996, the 202nd anniversary of George Washington's signing of the legislation authorizing *Constitution*'s construction. The renovation had taken forty-four months, cost twelve million dollars, and restored many important 1812-era features to the ship's structure. It also enabled Old Ironsides to do something it had not done in 116 years—sail under its own power. On 21 July 1997, the year of its 200th birthday, *Constitution* sailed once again in Massachusetts Bay. The suit of sails powering the ship was purchased with moneys raised in another pennies campaign by America's schoolchildren.

Old Ironsides
by Oliver Wendell Holmes

Ay, tear her tattered ensign down!
Long has it waved on high,
And many an eye has danced to see
That banner in the sky;
Beneath it rung the battle shout,
And burst the cannon's roar;—
The meteor of the ocean air
Shall sweep the clouds no more
Her deck, once red with heroes' blood,
Where knelt the vanquished foe,
When winds were hurrying o'er the flood,
And waves were white below,
No more shall feel the victor's tread,
Or know the conquered knee;—
The harpies of the shore shall pluck

The eagle of the sea!
Oh, better that her shattered hulk
Should sink beneath the wave;
Her thunders shook the mighty deep,
And there should be her grave;
Nail to the mast her holy flag,
Set every threadbare sail,
And give her to the god of storms,
The lightening and the gale!

Boston Daily Advertiser, 16 September 1830

CEB

Constitution's Spars, Rigging, and Sails

Wind and sail powered the American battle fleet in 1812. Then, a ship's power plant consisted of its spars, rigging, and sails, all of which were designed to capture the wind's motive force, propelling a vessel through the water. Four decades later, the last all sail-powered warship built by the U.S. Navy, the sloop of war *Constellation,* slipped down the ways at the Gosport Navy Yard. After that date, Navy-built ships would rely increasingly on mechanical propulsion systems powered by steam, and later, diesel, electrical, and atomic energy. As vessels employing steam technology entered the fleet in ever-greater numbers and wind-powered ones were struck from the Navy's rolls, ships like *Constitution* appeared increasingly obsolete, an aging reminder of a by-gone era when U.S. Sailors served in ships made of wood, rope, and canvas.

The text below offers a brief description of Old Ironsides' power plant—spars, rigging, and sails—which gave the frigate its distinctive profile and enabled it to sail on the ocean with speed, grace, and success for nearly half a century following its launch.

Spars

The primary function of *Constitution*'s spars is to carry sail. The frigate's largest spars are its three masts: mizzen, main, and fore. A formula based on the length and breadth of the ship's hull determined the size and positioning of these masts at construction. The mizzen, or aft-most, mast is mounted in a step (a large hollowed-out block) on the orlop deck and rises vertically through openings in the ship's decks to a height of 172 feet 6 inches. Both the main (center) and fore (forward) masts are seated in steps in the ship's keelson and stand 220 and 198 feet tall respectively. The masts themselves are not continuous sticks of wood, but rather a series of separate spars joined together in overlapping fashion to make a single mast, with each spar section smaller in length and thickness from that over which it is mounted. Each component of the mast structure bears its own name. The bottom-most spar of the mainmast, for example, is denominated the mainmast proper. The spars fixed above *Constitution*'s mainmast are known in succession as the main topmast, main topgallant mast, main royal mast, and main skysail mast. Platforms, called tops, connect each lower masthead to the topmast above it, while structures called crosstrees unite topmasts to topgallant masts.

Additional spars known as yards, booms, and gaffs act to carry *Constitution*'s sails. Yards carry the frigate's square-shaped sails and take their name from the mast section from which they hang. Thus the fore yard hangs from the foremast, the fore topsail yard from the fore topmast, and so on. The length of these yards can be increased by extending laterally spars known as studdingsail booms, from which additional sail can be set. Spars mounted to the mizzenmast (the gaff and spanker boom) and projecting from the bowsprit (the jib boom and flying jib boom) carry a number of *Constitution*'s fore and aft sails (sails aligned parallel to the keel).

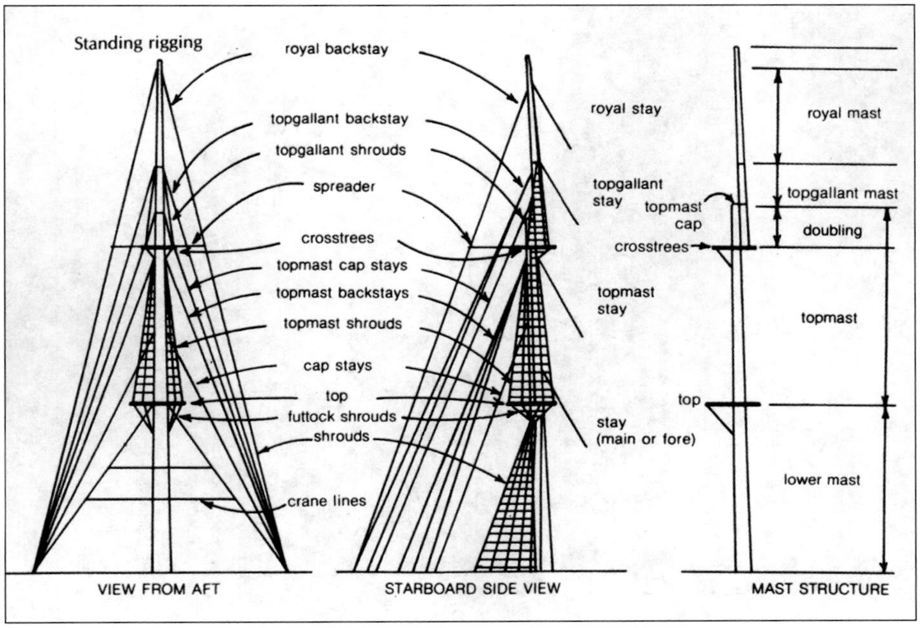

Profile views of the standing rigging and mast structure of a sailing warship. *Eagle Seamanship: A Manual for Square-Rigger Sailing*

A ship's spars are fashioned from a number of woods including fir, spruce, and white pine. The largest sections of *Constitution*'s original masts were fabricated from several pieces of white pine joined together to make a single spar. The resulting pieces, known as made masts, were actually stronger and more durable than if they had been built from a single length of timber. Masts mounted higher aloft in *Constitution* were crafted from individual sticks of wood and were known as pole masts. During its operational days, the ship carried a limited store of extra spars to replace or repair masts and yards damaged while under way. The carpenter and his mates were responsible for inspecting, maintaining, and repairing the frigate's spars.

Rigging

Constitution is fitted with two kinds of rigging, standing and running. The purpose of standing rigging is to support, stiffen, and stabilize the ship's masts. The most important components of this rigging are the stays, backstays, and shrouds. Stays are lines that run forward diagonally from one mast to another mast, the bowsprit, or jib boom. They help check rearward movement of the masts and serve to hang the majority of the frigate's fore and aft sails. Backstays are paired ropes leading aft and downward from the ship's masts to its sides. Backstays prevent the ship's masts from shifting forward and also provide a measure of support athwartships. Shrouds are ropes that brace the ship's masts sideways. Parallel sets of shrouds support each of the ship's lower masts and topmasts. Another set of shrouds supports the bowsprit. The lower mast shrouds extend from the channels along both sides of the ship to each masthead. The topmast shrouds stretch from each top to the crosstree above it. Ratlines, small ropes that cross the shrouds horizontally in ladderlike fashion, serve as steps for Sailors to climb aloft. A network of lanyards and blocks is used to tighten and maintain the proper tension in the ship's stays, backstays, and shrouds.

Constitution's running rigging consists of the ropes, chains, and blocks that operate and manage the ship's yards and sails. Tackles called jeers and halyards raise and lower the ship's yards vertically, while rope bridles (slings) and trusses (parrels) secure the yards to the center of their masts. Ropes extending diagonally from yardarms to mastheads, or lifts, also support the weight of the yards and keep them level. Lines extending aft from each yardarm called braces move the yards on a horizontal axis and are hauled on to trim the ship's sails. A rope known as a horse is stretched below each yard to provide footing for Sailors while they reef or furl sail.

A number of different lines control *Constitution*'s sails. The ship's square sails are furled by means of clew lines, buntlines, and slab lines. The first are attached to the sail's lower corners, or clews, while the remaining two are tied to the bottom edges or foot of each sail. Reef tackles fixed at the ends of each yard reduce sail by drawing up the sides, or leeches, of the sail toward the yardarm. Lines to trim *Constitution*'s square sails include the main and fore sail sheets, bowlines, and tacks. The first named lines run aft from clews, while the latter two run forward from leeches and clews respectively. Fore and aft sails set from the ship's stays are hoisted with halyard lines and trimmed with sheet and tack lines.

The amount of tackle required to rig a large man-of-war in the age of sail was considerable. It is estimated that about one thousand pulley blocks and forty miles of rope were required to rig a ship of the line properly. The boatswain and his mates were charged with keeping Old Ironsides' rigging in good order.

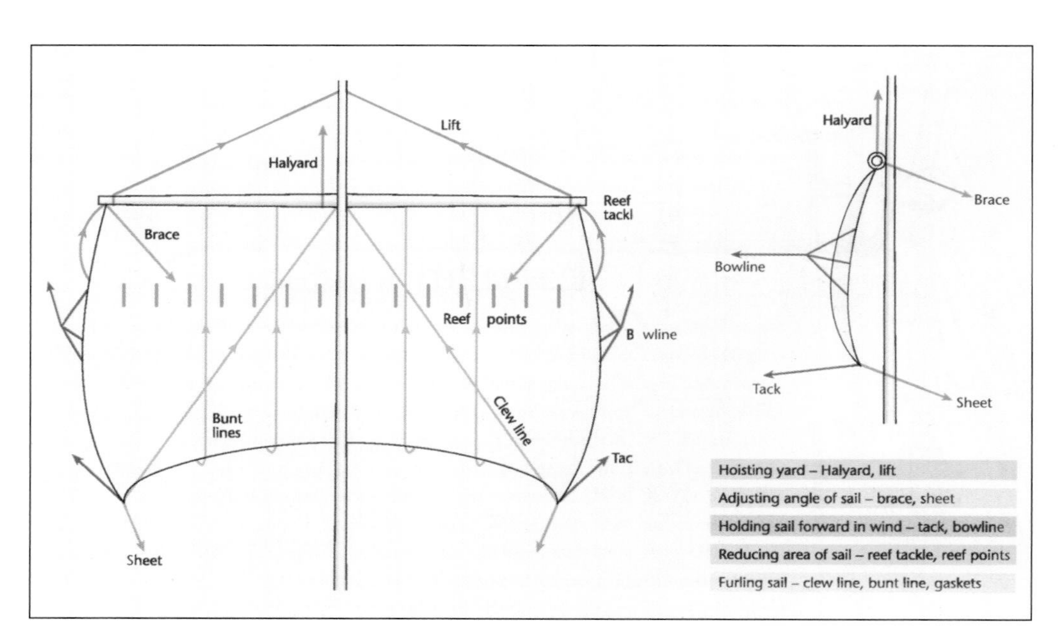

Diagram of the lines used to raise, lower, and trim square sails. *Courtesy of Conway Maritime Press, London, and taken from Jack Aubrey Commands*

Constitution's sail plan, 1817. *Naval Historical Center photograph*

Sails

Constitution carries two types of sail, square and fore and aft. The ship's square sails are not truly square, but are called so because they can be trimmed square—that is to say, at right angles with the ship's keel. Each sail takes its name from the yard from which it is set. Thus the main yard holds the mainsail or course, the main topsail yard holds the main topsail, and so on. The head, or top, of each sail is secured to its yard by short ropes called robands, while its foot is secured to the yard beneath it by means of sheet lines. The fore and main courses and the topsails have several rows of point-line sewn across them, called reef points, which are used to tie off and reduce sail. Light sails, known as studdingsails, are sometimes set in fair weather as a side extension of the ship's square sails. Except for the spanker and gaff topsail, the frigate's fore and aft sails are set on the ship's stays to which they are secured by rings called hanks. Like square sails, fore and aft sails take their name from the stays on which they are extended. For example, the main royal staysail is hung from the main royal stay.

Constitution carried up to forty different types of sail in its sail locker during the War of 1812. The number and type of sails set at any one time depended on several factors, the most important being wind and sea conditions. Generally speaking, sails were set from lower yards to higher ones, and struck in reverse order. Large sailing vessels required at least two to three knots of wind to gain steerageway and could carry full sail without hazard in winds of up to twenty-one knots. Heavy seas and high winds necessitated reducing sail. Sails were also reduced when going into battle to minimize damage to canvas from shot, shrapnel, etc., as well as to lessen the number of hands aloft. The topsails were the most important sails in *Constitution*. They served well in a variety of sea conditions and were essential to the ship's maneuverability under way. The main topsail was the largest piece of canvas Old Ironsides carried. The ship's highest sails, royal and sky sails, were set in light winds. Its staysails assisted in stormy weather, while its headsails (those set forward of the foremast), spanker, and gaff topsail helped steer the ship.

The canvas for *Constitution*'s 1812 sails was manufactured from cloth made of hemp. This fabric was graded according to thickness with number one cloth being the thickest and heaviest and number eight being the lightest. Heavier sailcloth was used to make the ship's lower sails and storm sails, while lighter cloth was used for its upper and studdingsails. High quality cloth called Ravens-duck or "ravens" and Russia duck were also used in the ship's light sails. *Constitution*'s original suit of sails was made from Connecticut-grown hemp. The sailmaker and his mates oversaw the making, inspection, and repair of the frigate's sails.

CEB

History of the Turnaround Cruises and OPSAIL 200

The first mention of a turnaround cruise for *Constitution* was in the 1950s when the ship was infrequently turned around in its berth to weather the hull evenly. This was done on no fixed schedule, perhaps once or twice a year. Several traditions regarding the turnaround have evolved since then. On 20 November 1959 Old Ironsides' commander, Lieutenant Edward J. Melanson, invited male guests to participate in the cruise. During the 1960s *Constitution* hosted some four hundred men for its annual tow down Boston Harbor and back. The male-only cruises ceased after 1971, when a female stowaway, Ruby Litinsky, a journalist from Peabody, Massachusetts, exposed the sexist practice. Today, a lottery system dispenses tickets for the turnaround and the other popular towed cruises that occur several times each year. Another 1960s innovation, that of participants receiving a certificate for completing the cruise "with intrepidity," continues to this day.

Constitution celebrated America's bicentennial on 17 June 1976 with youngsters from thirty-three different nations as guests. Three weeks later while the frigate hosted the "Tall Ships" parade, hundreds of thousands of people watched and cheered from every possible vantage point as the frigate, its guns firing at one-minute intervals, led the fleet through Boston Harbor.

Today's customs of scheduling a turnaround cruise for Independence Day and of firing a twenty-one-gun salute to the nation and a nineteen-gun salute to the Commonwealth of Massachusetts date to 1977. In 2005 the ship invited all former captains of the ship to participate in the turnaround ceremonies.

The turnaround cruise has evolved since the 1950s beyond its original utilitarian function of weathering the ship to having both ceremonial and educational roles. Several cruises from June through August celebrate these themes: Bunker Hill, Independence Day, USS Constitution Museum, and Old Ironsides Across the Nation. In addition, since 1997 the Navy's Leadership Continuum Program has scheduled two cruises each year to train hundreds of Sailors in handling sails. USS *Constitution* commemorated its own bicentennial by cruising under sail for the first time in 116 years on 21 July 1997 in Operation Sail 200 (OPSAIL 200). *Constitution* had recently completed a four-year overhaul to restore it to original strength. In-depth structural tests and evaluations of the ship and a lengthy training program for the crew prepared the ship for the historic voyage.

Under the command of Commander Michael C. Beck and Lieutenant Commander Clair V. Bloom, *Constitution*'s first female officer and its first female XO, the ship was towed from Boston to Marblehead, Massachusetts, seventeen miles to the north. The latter city has historic significance for Old Ironsides, as it sought refuge there from British warships during the War of 1812 and the ship visited that port in 1931 during its three-coast tour. The return trip from Marblehead to Boston marked the first time since 1881 that the world's oldest commissioned warship afloat had sailed under wind power, using the same six-sail configuration that it normally used in battle. USS Constitution Museum, a private, nonprofit, educational institution dedicated to preserving the heritage of the historic frigate, coordinated a nationwide "Ironsides Pennies Campaign" to purchase new sails for the frigate. School children from all fifty states contributed their pennies to the cause.

Accompanied by a fleet of yachts and sailboats, and under the escort of two active duty Navy warships, the guided missile destroyer *Ramage* (DDG 61) and the

Constitution hosted more than seventy Medal of Honor recipients for a turnaround cruise on 30 September 2006. *U.S. Air Force photograph*

guided missile frigate *Halyburton* (FFG 40), *Constitution* sailed for one hour in Massachusetts Bay during OPSAIL 200. The Navy's flight demonstration team, the Blue Angels, rendered *Constitution* full honors while passing overhead. Thousands of visitors cross the decks of Old Ironsides each year in homage to the ship and Sailors who served America in war and peace.

CFH

On 21 July 1997 *Constitution* celebrated the bicentennial of its launching, accompanied by a flotilla of vessels, during its memorable sail from Marblehead, Massachusetts, to its berth in Boston. *Naval Historical Center photograph*

Constitution's Uncommon Connections: Presidents, Potentates, and a Pope

From its very origins, the frigate *Constitution* has had connections with heads of state and other world leaders.

- President George Washington authorized the frigate and assigned its name.
- President Andrew Jackson's likeness adorned its bow as a figurehead for about forty years, from 1834 to the early 1870s.
- Pope Pius IX visited the ship on 1 August 1849, becoming the first pope to visit U.S. territory.
- President Ulysses S. Grant, in 1869, was the first U.S. president to tread *Constitution*'s decks.
- President Herbert Hoover is the last U.S. president in office to come on board (1931 and 1932), at least until another sitting president should make a visit.
- Queen Elizabeth II and Prince Philip of the United Kingdom visited the frigate in Boston, Massachusetts, during the bicentennial of the United States in 1976.

President Jackson Beheaded

The story of *Constitution*'s Andrew Jackson figurehead is one of the more amusing episodes in the frigate's long and distinguished history. Soldier, Democratic politician, and president from 1829 to 1837, Jackson was among the most controversial men ever to hold the presidency of the United States. Much admired in the West and South, he was heartily despised by many in New England, especially in the Whig stronghold of Boston, Massachusetts.

Constitution carried this billethead bearing the figure of a winged dragon during the War of 1812. The frigate's original figurehead of a lion-skinned Hercules holding the Constitution (document) was removed after being damaged in a collision with *President* in 1804. *Naval Historical Center photograph*

In 1833 the Jackson administration appointed Navy Captain Jesse Duncan Elliott commandant of the Charlestown (Massachusetts) Navy Yard, where *Constitution* was receiving its first major overhaul. To register his admiration of Jackson, and perhaps to curry further favor with the administration, Elliott contracted to have a new figurehead, a full figure from head to foot in the likeness of President Jackson, carved for the frigate. The original figurehead, in the form of Hercules holding in one hand a fasces (bound sticks representing the strength of union) and in the other a roll of parchment representing the Constitution of the United States, had been destroyed in a collision with USS *President* in the Mediterranean in 1804. A plain billethead had adorned *Constitution*'s prow since that time. Elliott engaged a young wood sculptor named Laban S. Beecher to carve the new figurehead.

Word leaked out and someone circulated handbills denouncing what Bostonians viewed as the intended desecration of the frigate they had embraced as their own. After Beecher received threats of violence, Elliott made room for the artist's studio in the navy yard, where he would be protected. When the Board of Navy Commissioners in Washington learned about the carving, they informed Elliott that the department had long since decided that only ships of the line were to have figureheads. Nevertheless, since the project was nearly finished, the board gave Elliott the choice of proceeding as he had planned or saving the figurehead for a future ship of the line. Elliott chose the former course of action and had "Old Hickory's" likeness attached to *Constitution*'s prow in April 1834.

On the morning of 3 July, Elliott walked out to find the figurehead mutilated. Someone had sawed off the president's head! Elliott draped a cover over the figure until he got a new head carved—this time in New York City—and offered a reward for information leading to the arrest of the saw-wielding culprit. The decapitated head later turned up on the desk of Secretary of the Navy Mahlon Dickerson, and the story emerged that, on a bet, Samuel Worthington Dewey, a young merchant shipmaster, concealed by the darkness of night and a heavy rain, had rowed out to the wharf where *Constitution* was moored, hoisted himself up the bow, and cut off the offending head.

In 1848, the original Jackson figure (and head number two) was replaced with a new version by J. D. and W. H. Fowle, carvers of Boston—apparently after a dozen or so years the passionate anti-Jackson feelings of 1834 had cooled. The new figurehead adorned *Constitution* until a billethead replaced it, restoring the ship to a profile closer to what it had during the War of 1812, in time for the observances of the nation's centennial in 1876.

Papal Visit Creates Tempest in a Teapot

During the last years of the 1840s *Constitution* served in the Mediterranean Squadron, protecting U.S. citizens from violent political revolutions convulsing much of Europe, including the various independent states of the Italian peninsula. In 1849 the American chargé d'affaires in Naples and Rome suggested to *Constitution*'s captain, John Gwinn, that the frigate stop at Gaeta, a city on the west coast of Italy roughly seventy-five miles southeast of Rome, so that Pope Pius IX and Ferdinand II, King of the Two Sicilies, could pay the ship a visit. In July Gwinn brought the proposal to the squadron's newly arrived commodore, Charles W. Morgan. Morgan promptly and vehemently rejected the idea. In order to maintain the United States' policy of neutrality, the commodore insisted on avoiding any appearance of taking sides in the revolutionary contests raging in both the Papal States and Sicily. Morgan ordered *Constitution* to the coast of northern Italy, by way of Messina and Sardinia.

Disregarding Morgan's orders, Gwinn followed the instructions of his passenger, John Rowan, U.S. chargé d'affaires in Naples, and sailed *Constitution* directly to Gaeta. There, on 1 August 1849, *Constitution* entertained King Ferdinand and Pope Pius IX. In coming on board *Constitution* the latter visitor became the first pope to step foot on U.S. territory. The king explored the entire ship with interest, and the pope gave a benediction to the eighty Catholics in the crew, to whom he later sent rosaries. During the visit, *Constitution*'s surgeon treated the pope for seasickness. When the dignitaries departed, the crew manned the yards and fired a twenty-one-gun salute.

Aghast at his subordinate's disobedience of orders and violation of American neutrality, Commodore Morgan recommended that Gwinn be transferred to a station less desirable than the Mediterranean. But at Palermo on 4 September, before any such mark of disapprobation could be imposed, Gwinn died of natural causes.

MJC

Disregarding squadron orders to remain neutral among warring political factions in mid-nineteenth century Italy, *Constitution*'s Captain John Gwinn entertained both Pope Pius IX, shown here, and King Ferdinand II when the frigate visited the western coast of Italy in 1849. *available on the Web*

It now appeared that we must be taken, and that our Escape was impossible, four heavy Ships nearly within Gun Shot, and coming up fast, and not the least hope of a breeze, to give us a chance of getting off by out sailing them.

Part IV: **Appendices**

The documents that follow adhere as closely as possible to the originals in spelling, capitalization, punctuation, and abbreviations. For clarity's sake, dates for Constitution*'s log book entries have been converted from sea time to civil time. Until 1848, U.S. Navy log books reckoned the day in sea time which ran from one second past noon of one day to noon of the next. After that date, Navy logs measured the day in civil time, that is, from one second past midnight to midnight of the next day.*

An Act to Provide a Naval Armament, 27 March 1794

WHEREAS the depredations committed by the Algerine corsairs on the commerce of the United States render it necessary that a naval force should be provided for its protection:

SECTION 1. *Be it therefore enacted by the Senate and House of Representatives of the United States of America in Congress assembled,* That the President of the United States be authorized to provide, by purchase or otherwise, equip and employ four ships to carry forty-four guns each, and two ships to carry thirty-six guns each.

SEC. 2. *And be it further enacted,* That there shall be employed on board each of the said ships of forty-four guns, one captain, four lieutenants, one lieutenant of marines, one chaplain, one surgeon, and two surgeon's mates; and in each of the ships of thirty-six guns, one captain, three lieutenants, one lieutenant of marines, one surgeon, and one surgeon's mate, who shall be appointed and commissioned in like manner as other officers of the United States are.

SEC. 3. *And be it further enacted,* That there shall be employed, in each of the said ships, the following warrant officers, who shall be appointed by the President of the United States, to wit: One sailing-master, one purser, one boatswain, one gunner, one sail-maker, one carpenter, and eight midshipmen; and the following petty officers, who shall be appointed by the captains of the ships, respectively, in which they are to be employed, viz: two master's mates, one captain's clerk, two boatswain's mates, one cockswain, one sail-maker's mate, two gunner's mates, one yeoman of the gun room, nine quarter-gunners, (and for the four larger ships two additional quarter-gunners,) two carpenter's mates, one armourer, one steward, one cooper, one master-at-arms, and one cook.

SEC. 4. *And be it further enacted,* That the crews of each of the said ships of forty-four guns, shall consist of one hundred and fifty seamen, one hundred and three midshipmen and ordinary seamen, one sergeant, one corporal, one drum, one fife, and fifty marines; and that the crews of each of the said ships of thirty-six guns shall consist of one hundred and thirty able seamen and midshipmen, ninety ordinary seamen, one sergeant, two corporals, one drum, one fife, and forty marines, over and above the officers herein before mentioned.

SEC. 5. *And be it further enacted,* That the President of the United States be, and he is hereby empowered, to provide, by purchase or otherwise, in lieu of the said six ships, a naval force not exceeding, in the whole, that by this act directed, so that no ship thus provided shall carry less than thirty-two guns; or he may so provide any proportion thereof, which, in his discretion, he may think proper.

SEC. 6. *And be it further enacted,* That the pay and subsistence of the respective commissioned and warrant officers be as follows:—A captain, seventy-five dollars per month, and six rations per day;—a lieutenant, forty dollars per month, and three rations per day;—a lieutenant of marines, twenty-six dollars per month, and two rations per day;—a chaplain, forty dollars per month, and two rations per day;—a sailing-master, forty dollars per month, and two rations per day;—a surgeon, fifty dollars per month, and two rations per day;—a surgeon's mate, thirty dollars per month, and two rations per day;—a purser, forty dollars per month, and two rations per day;—a boatswain, fourteen dollars per month, and two rations per day;—a gunner, fourteen dollars per month, and two rations per day;—a sail-maker, fourteen dollars per month, and two rations per day;—a carpenter, fourteen dollars per month, and two rations per day.

SEC. 7. *And be it further enacted,* That the pay to be allowed to the petty officers, midshipmen, seamen, ordinary seamen and marines, shall be fixed by the President of the United States: *Provided,* That the whole sum to be given for the whole pay aforesaid, shall not exceed twenty-seven thousand dollars per month, and that each of the said persons shall be entitled to one ration per day.

SEC. 8. *And be it further enacted,* That the ration shall consist of, as follows: Sunday, one pound of bread, one pound and a half of beef, and half a pint of rice:—Monday, one pound of bread, one pound of pork, half a pint of peas or beans, and four ounces of cheese:—Tuesday, one pound of bread, one pound and a half of beef, and one pound of potatoes or turnips, and pudding: Wednesday, one pound of bread, two ounces of butter, or in lieu thereof, six ounces of molasses, four ounces of cheese, and half a pint of rice:—Thursday, one pound of bread, one pound of pork, and half a pint of peas or beans:—Friday, one pound of bread, one pound of salt fish, two ounces of butter or one gill of oil, and one pound of potatoes:—Saturday, one pound of bread, one pound of pork, half a pint of peas or beans, and four ounces of cheese.—And there shall also be allowed, one half pint of distilled spirits per day, or, in lieu thereof, one quart of beer per day, to each ration.

Sec. 9. *Provided always, and be it further enacted,* That if a peace shall take place between the United States and the Regency of Algiers, that no farther proceeding be had under this act.

Approved, March 27, 1794.

Statutes at Large of the United States of America, 1789–1845, 8 vols. (Boston: Little, Brown and Company, 1846–67), 1:350–51. This act was Chapter XII of the statutes passed during the first session of the third Congress.

Captain Isaac Hull To Secretary of the Navy Paul Hamilton

US. Frigate *Constitution*
At Sea July 21st 1812.

Sir,

In pursuance of your orders of the 3d. inst. I left annapolis on the 5th. inst and the Capes on the 12th. of which I advised you by the Pilot that brought the Ship to sea

For several days after we got out the wind was light, and ahead which with a strong Southerly current prevented our making much way to the Northward On the 17th. at 2 PM. being in 22 fathoms water off Egg harbour four sail of Ships were discovered from the Mast Head to the Northward and in shore of us; apparently Ships of War The wind being very light all sail was made in chase of them, to ascertain whether they were Enemy's Ships, or our Squadron having got out of New York waiting the arrival of the *Constitution*, the latter of which, I had reason to believe was the case.;

At 4 in the afternoon a Ship was seen from the Mast head bearing about NE. Standing for us under all sail, which she continued to do until Sundown at which time, she was too far off to distinguish signals and the Ships in Shore, only to be seen from the Tops, they were standing off to the Southward, and Eastward. As we could not ascertain before dark, what the Ship in the offing was, I determined to stand for her and get near enough to make the night signal. At 10 in the Evening being within Six or Eight miles of the Strange sail, the Private Signal was made, and kept up nearly one hour, but finding she could not answer it, I concluded she, and the Ships in shore were Enemy. I immediately hauled off to the Southward, and Eastward, and made all sail, having determined to lay off till day light, to see what they were. The Ship that we had been chasing, hauled off after us shewing a light, and occasionally making signals, supposed to be for the Ships in shore.

18th. At day light, or a little before it was quite light, saw two sail under our Lee, which proved to be Frigates of the Enemies. One Frigate astern within about five or six miles, and a Line of Battle Ship, a Frigate, a Brig, and Schooner, about ten or twelve miles directly astern all in chase of us, with a fine breeze, and coming up very fast it being nearly calm where we were. Soon after Sunrise the wind entirely left us, and the Ship would not steer but fell round off with her head towards the two Ships under our lee.

The Boats were instantly hoisted out, and sent ahead to tow the Ships head round, and to endeavour to get her farther from the Enemy, being then within five miles of three heavy Frigates. The Boats of the Enemy were got out, and sent ahead to tow, which with the light air that remained with them, they came up very fast. Finding the Enemy coming fast up, and but little chance of escaping from them; I ordered two of the Guns on the Gun Deck, ran out at the Cabbin windows for stern Guns on the gun deck, and hoisted one of the 24 Pounders off the Gun deck, and run that, with the Fore Castle Gun, an Eighteen pounder, out at the Ports on the Quarter Deck, and cleared the Ship for Action, being determined they should not get her, without resistance on our part, notwithstanding their force, and the situation we were placed in.

At about 7 in the Morning the Ship nearest us, approaching with Gun Shot, and directly astern, I ordered one of the stern Guns fired to see if we could reach her, to endeavour to disable her masts, found the Shot fell a little Short, would not fire any more.

At 8 four of the Enemy's Ships nearly within Gun Shot; some of them having six or eight boats ahead towing, with all their Oars, and Sweeps out to row them up with us, which they were fast doeing, It now appeared that we must be taken, and that our Escape was impossible, four heavy Ships nearly within Gun Shot, and coming up fast, and not the least hope of a breeze, to give us a chance of getting off by out sailing them.

In this Situation finding ourselves in only twenty four fathoms water (by the suggestion of that valuable officer Lieutenant Morris) I determined to try and warp the Ship ahead, by carrying out anchors and warp her up to them, Three or four hundred fathoms of rope was instantly got up, and two anchors got ready and sent ahead, by which means we began to gain ahead of the Enemy, They however soon saw our Boats carrying out the anchors, and adopted the same plan, under very advantageous circumstances, as all the Boats, from the Ship furthermost off were sent to Tow, and Warp up those nearest to us, by which means they again came up, So that at 9 the Ship nearest us began firing her bow guns, which we instantly returned by our Stern guns in the cabbin, and on the Quarter Deck; All the Shot from the Enemy fell short, but we have reason to believe that some of ours went on board her, as we could not see them strike the Water.

Soon after 9 a Second Frigate passed under our lee, and opened her Broadside, but finding her shot fall short, discontinued her fire, but continued as did all the rest of them to make every possible exertion to get up with us. From 9 to 12 all hands were employed in warping the Ship ahead, and in starting some of the water in the main Hold, to lighten her, which with the help of a light air, we rather gained of the Enemy, or at least hold our own. About 2 in the afternoon, all the Boats from the Line of Battle Ship, and some of the Frigates, were sent to the Frigate nearest to us, to endeavour to tow her up, but a light breeze sprung up, which enabled us to hold way with her notwithstanding they had Eight or Ten Boats ahead, and all her sails furled to tow her to windward. The wind continued light until 11 at night, and the Boats were kept ahead towing, and warping to keep out of the reach of the Enemy, Three of the Frigates being very near us. At 11 we got a light breeze from the Southward, the boats came along side, and were hoisted up, the Ship having too much way to keep them ahead, The Enemy still in chase, and very near.

19th. At daylight passed within gun shot of one of the Frigates, but she did not fire on us, perhaps for fear of becalming her as the wind was light soon after passing us, she tacked, and stood after us, At this time Six sail were in Sight under all sail after us.

At 9 in the morning saw a Strange sail on our Weather Beam, supposed to be an American Merchant ship, the instant the Frigate, nearest us saw her she hoisted American colours, as did all the Squadron in hopes to decoy her down, I immediately hoisted English colours, that she might not be deceived, she soon hauled her wind, and it is to be hoped made her escape. All this day the Wind increased gradually and we gained on the Enemy, in the course of the day Six or Eight miles, they however continued chasing us all night under a press of sail.

20th. At day light in the Morning only three of them could be seen from the Mast head, the nearest of which, was about 12 miles off directly astern. All hands were set at work wetting the Sails, from the Royals down, with the Engine, and Fire buckets, and we soon found that we left the Enemy very fast. At ¼ past 8 the Enemy finding that they were fast dropping astern, gave over chase, and hauled their wind to the Northward, probably for the Station off New York. At ½ past 8 Saw a sail ahead gave chase after her under all sail; At 9 Saw another Strange sail under our Lee Bow, we soon spoke the first sail, discovered and found her to be an American Brig from St. Domingo bound Portland, I directed the Captain how to steer to avoid the Enemy, and made sail for the vessel to leeward, on coming up with her, She proved to be an American Brig from St. Bartholemews, bound to Philadelphia, but on being informed of War he bore up for Charleston, S.C.

Finding the Ship so far to the Southward, and Eastward, and the Enemy's Squadron stationed off New York, which would make it impossible for the Ship to get in there. I determined to make for Boston to receive your further orders, and I hope that my having done so will meet your approbation. My wish to explain to you as clearly as possible why your orders, have not been executed, and the length of time the Enemy were in chase of us with various other circumstances, has caused me to make this communication much longer than I would have wished, yet I cannot (in justice to the brave Officers, and crew, under my Command) close it without expressing to you the confidence I have in them, and assuring you that their conduct whilst under the Guns of the Enemy was such as might have been expected from American Officers and Seamen, I have the Honour to be, with very great Respect, Sir, Your Obt. Hbl Servt.

Isaac Hull

National Archives and Records Administration, Washington, DC, Record Group 45, Captains' Letters (Microfilm 125, Reel 24, Letter No. 127). Because Hull's description of the events above is based on the dates and times recorded in *Constitution*'s logbook, he gave them as they occurred in sea time. The actual dates are 16–19 July 1812.

USS *Constitution* Logbook, 16–19 July 1812

Remarks on Thursday July 16th 1812

… [*At noon*] Commences with clear weather and fresh breezes from the Northward and Eastward, at 1 ½ PM Sounded in 22 fms of water at 2 PM 4 Sail of Vessels in Sight, turned the 2d. and 3d. Reefs out of the Topsails, and sett Top Gallant Sails, at ½ past 2 PM sett the Royals, at 3 PM sounded in 18 ½ fms of water, at ¼ past 3 PM tacked to the Eastd., at 4 PM a Ship in Sight bearing N.E standing down for us and three Ships, and a Brig N.N.W on the Starboard Tack J[*ohn*]. T. S[*hubrick*]

From 4 to 6 PM light airs from the Northward at ¼ past 5 PM hauled up the Main Sail, at 6 PM the Single Ship bearing E. N. E. A[*lexander*]. S. W[*adsworth*]

At ¼ past 6 PM got a light breeze from the Southward; and Eastward, wore ship, and stood towards the above Sail keeping her a little off the Larboard Bow, sett Top Gallant, and Royal Staysails, and Starboard Top Gallant Studding Sails at ½ past 7 beat to quarters, and cleared ship for Action, at 8 PM light Airs, Coming up with the above Ship very slow. Geo. C Reed.

At 10 PM hauled down the Stay Sails and hauled up the spanker, at ½ past 10 PM made the private Signal of the day, at ¼ past 11 PM hauled down the Signals, not having been answered by the above Ship, and made Sail by the wind, with starboard Tacks on board

Remarks on Friday July 17th 1812

from 12 to 4 AM, light airs from the Southward, and westward, and cloudy, at 4 AM the Ship made a Signal (a rocket, and two Guns) at day light discovered three Sail off the Larboard Quarter bearing N. E. and three Sail a Stern. J.T.S.

At 5 Am discovered another Sail astern making 2 Frigates off our Lee quarter, and 2 Frigates, and one Ship of the Line, one Brig, and one Schooner a Stern at ¼ past 5 Am, it being calm, and the Ship having no Steerage way hoisted out the First Cutter and got the Boats a head to tow Ship's head round to the Southward, got a 24 pounder up off the Gun deck, for a Stern Gun and the Fore Castle Gun aft, Cut away the Taffierail to give them room, and run two Guns out of the Cabin windows, at 6 PM [*a.m.*] got the Ship's head round to the Southward, and Sett Top Gallant Stud'g. Sails, and Staysails, one of the Frigates Firing at us, at ½ past 6 Am Sounded in 26 fms. of water, at 7 Am got out a Kedge and warped the Ship a head, at ½ past 7 AM, hoisted the Colours, and fired one Gun, at the Ship astern, at 8 Am calm, employed warping, and towing the Ship a head, The other Ships having a light air gaining on us, with their boats a head, and one of them using their Sweeps, at 9 Am the above Ship in Close chace of us, and nearest Frigate gaining on us at 9 minutes past 9 Am a light breeze Sprung up from the Southward, braced up by the wind on the Larboard Tack, when the above Frigate commenced firing but her shot did not reach us, got the Boats alongside, run two of them up, at 10 Am started about 2,335 Gallons of water, and pumped it out, almost calm, manned the First Cutter to tow Ship, Six sail of the Enemies Ships off the Starboard Beam, and quarter, perceived that the nearest Frigate had got all the Boats from the other Ships to tow her towards us, From 10 Am to Meridian employed warping, and towing, all sail made by the wind, one of the above Ships Coming up, apparently having all the Boats from the other Ships light airs, and cloudy. Geo. C. Reed. Latie. Obserd. 39° 15' N

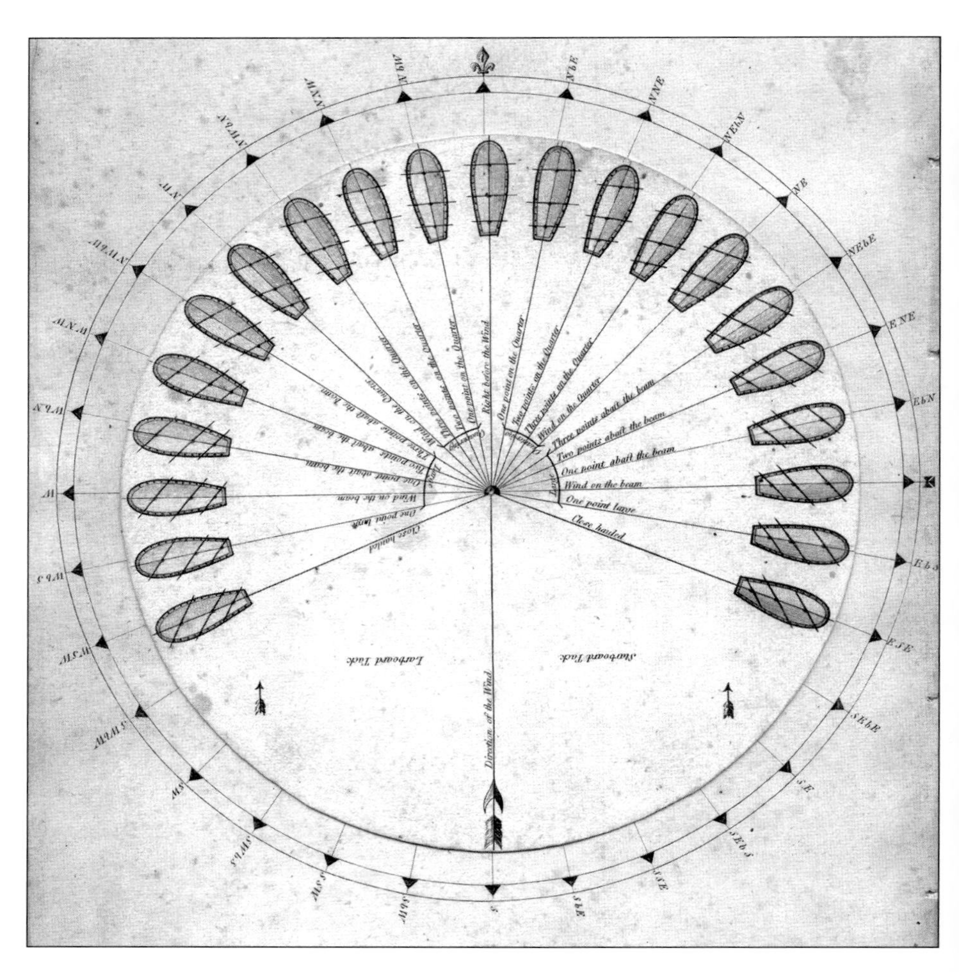

This eighteenth-century compass was used to teach aspiring ship captains the art of navigation. Each of the compass's thirty-two points is denominated, as are points of sailing relative to the wind. Square-rigged vessels could sail no closer than six points off the wind when working to windward. *Elements and Practice of Rigging and Seamanship, Vol. 2, 1794*

Commences with light airs from the Southward, and Eastward, attended with Calms, at ¼ after Meridian sent the First Cutter, and Green Cutter a head to tow Ship at ¼ before 1 PM a strange Sail, discovered two points abaft off the Lee Beam, the 4 Frigates One point off the Starbod. quarter Line of Battle Ship, Brig and, Schooner off the Lee Beam, at 7 minutes before 2 P.M, the chasing Frigates, Commenced firing their Bow Chace Guns, we returned them with our Stern Chase Guns, at ½ past 3 PM still chased by the above Ships, one of them which is nearly within Gun shot, at 7 PM observed the Enemies Ships towing with their Boats, Lowered down the First Cutter, Green Cutter, and Gig, and sent them a head to tow Ship, steering S W ½ W light airs inclinable to Calms, at ½ past 7 PM Sounded in 24 fms of water at 8 PM light airs from the Southward, and Eastward the 1st. and 5th. Cutters, and Gig a head towing ship the Enemies Ships in the same position as at ½ past 7 PM, From 8 to 9 PM light Airs and cloudy, Enemies Ships still in chase of our Boats a head towing Ship, at 7 minutes before 11 PM a breeze sprung up from the Southward, boats came along side hoisted up the Gig and Green Cutter, and sett the Fore top Mast staysail, and Main Top Gallant Studding sail, at Midnight moderate breezes, and cloudy, sounded in 26 fms of water the Enemies Ship Still in Chase. Geo. C Reed,

Remarks on board on Saturday July 18th. 1812

at 2 AM sounded in 23 fms of water, discovered one of the Ships off the Lee Beam, at ½ past 2 AM took in the Steering sails, at day light Four Frigates in Sight, three off the Lee quarter, and one off the Lee Beam from 2 to 3 Miles distant, at 4 AM Six Sail in Sight from off the deck, hauled down the Fore top mast Stay sail very light breezes. B[*eekman*]. V. H[*offman*],

At 4 hours 20 minutes Am, Tacked ship to the Eastward, at 5 Am passed about Gun Shot distance to windward of one of the Frigates, hoisted in the First Cutter, Ten sail in Sight from the Mast head. J.T.S,

At 8 Am hauled down the Middle Staysail, at 9 Am fitted and sett Fore and Main Sky sails saw a Ship to windward, supposed to be an American Merchantman Standing toward us, the Frigate astern hoisted American Colours, as a decoy we immediately hoisted English, Colours, set Royal Studding Sails, and Booms fitted, and Shifted the Starboard Fore top mast studding sail boom which was sprung at 10 ¼ am sounded in 25 fms of water, fine Grey sand, and broken shells, at 11 Am took in Stay Sails at Meridian moderate breezes, and pleasant weather, rather leaving the Frigates in chase, the headmost Frigate to Leeward bearing nearly N. by W, 4 or 5 Miles distant, the nearest Frigate W. N. W direct in our wake distant about 3 ½ miles the Line of Battle Ship N. by W. ½. W on the Larboard Tack hull down, Two Frigates off our Lee quarter N. N. W ½. W, and N. W by N about 5 Miles distant, and a Brig bearing about N. by W, Observed Latitude 38° 47' North which from that, and the Soundings got at ¼ past 10 AM allowing for the distance since run gives our Longe. about 73° 53' West from which we date our Departure. A.S.W

Commences with fresh breezes from the Southward, and pleasant, at 1 PM hauled down the Royal Stay sails, and sett the Middle Stay sail, at 2 PM got shifting Backstays on the Top Gallant Masts, and sett them well up, took in the Gaff Topsail, and Mizen Top Gallant staysail, at ½ past 2 PM sett the Mizen Top Gallant, and Main Royal Staysails, and Main Skysails, at 4 pm a moderate breeze from the S. S. W, and cloudy four sail of the Enemy still in chase the nearest about 6 Miles of[*f*] bearing N. N. W ¼ W and one off the weather Quarter W by N. ½ N. Geo. C Reed.

From 4 to 6 PM moderate breezes and Cloudy, at 5 PM sett Top Gallant, and Fore Topmast Studding Sails ¼ before 6 PM took in Top Gallant Studding sails, and Fore Top mast Studding sail, at ½ past 6 PM took in the Royals, at ¼ before 7 PM a heavy squall of wind, and rain took in staysails Top Gallant sails, and flying Jib, Sett the Fore top mast staysail, and took the 2d. reef in the Mizin Topsail, and one reef in the Spanker, at 7 PM sett Top Gallant sails, and Main Top Mast Stay sail, at ½ past PM the Leewardmost Ship N. N. W ½ W, and the weathermost Ship N. W. By W. ¾. W, the other two more astern and hull down, Sett Middle, and Top Gallant Staysails, at ¾ past 7 PM Sett Royals Flying Jib, and Mizen Top Gallant, staysail, and turned the reef out of the Spanker, Sent the Skysail yards down in the Top, at ¼ past 8 PM wind light, Sett the Starboard Top Gallant Studding Sails, at ½ past 9 PM the wind hauled round to the Southward, and westward, sett Starboard Lower steering, and Top mast, and Royal Studding Sails Sky sails, and Gaff Top sails, rounded in the weather braces, at ½ past 10 PM, the wind backed round again took in the Lower Steering sail, and braced up, heard two Guns from the Enemies Ships off the Lee quarter, at 11 PM could just discover, the weather Ship to have got in our wake, at Midnight moderate breezes, and pleasant took in the Royal Studding Sails. Geo. C Reed

Remarks on board Sunday July 19th 1812

From Midnight to 4 AM moderate breezes and cloudy, at 1 Am sett the Skysails, at ¼ before 2 AM got a pull of the weather braces and sett the Lower Steering sail, at 3 AM sett the Main Topmast Studding Sail, at ¼ past 4 AM hauled up to S. E by S. 4 Sail in Sight astern, the westernmost ship bearing N. W ½ W the 2d. N. W and the others N.W. by N. northerly, all of them hull down, at ½ past 6 AM more moderate, employed wetting the Sails aloft, at 8 AM moderate breezes and pleasant 4 Ships still in sight chacing of us, the nearest, and weathermost ship having her lower yards under the same ship bearing N. W ¼ N. and the Leeward Ship N by W. A.S.W.

At ¼ past 8 AM all the Ships in chase stood to the Northward and Eastward… .

National Archives and Records Administration, Washington, DC, Record Group 24, Logbooks and Journals of the USS *Constitution*, 1798–1934 (Microfilm 1030, Reel 1). Heading at top of page reads: "The United S. Frigate *Constitution* Isaac Hull Esqr. Commanr." Although this event took place on 16–19 July 1812, the logbook records it in sea time under the dates 17–19 July.

Captain Isaac Hull to Secretary of the Navy Paul Hamilton

US Frigate *Constitution*
Off Boston Light August 28th. 1812

Sir,

I have the Honour to inform you that on the 19th inst. at 2 PM being in Lattitude 41°.42' Longitude 55°.48' with the wind from the Northward, and the *Constitution* under my command steering to the S.SW. a sail was discovered from the Mast head bearing E by S. or E.SE. but at such a distance that we could not make out what she was. All sail was immediately made in chace, and we soon found we came fast up with the chace, so that at 3 PM. we could make her out to be a Ship on the Starboard tack close by the wind under easy sail. At ½ past 3 PM. Closing very fast with the chace could see that she was a large Frigate, At ¾ past 3 the chace backed her maintopsail, and lay by on the Starboard tack; I immediately ordered the light sails taken in, and the Royal Yards sent down, took two reefs in the topsails, hauled up the foresail, and mainsail and see all clear for action, after all was clear the Ship was ordered to be kept away for the Enemy, on hearing of which the Gallant crew gave three cheers, and requested to be laid close alongside the chace. As we bore up she hoisted an English Ensign at the Mizen Gaff, another in the Mizen Shrouds, and a Jack at the Fore, and MizentopGallant mast heads. At 5 minutes past 5 P M. as we were running down on her weather quarter She fired a Broadside, but without effect the Shot all falling short, she then wore and gave us a broadside from her Larboard Guns, two of which Shot Struck us but without doeing any injury. At this time finding we were within gunshot, I ordered the Ensign hoisted at the Mizen Peak, and a Jack at the Fore and MizentopGallant mast head, and a Jack bent ready for hoisting at the Main, the Enemy continued wearing, and manoeuvering for about ¾ of an hour, to get the wind of us. at length finding that she could not, she bore up to bring the wind, on the quarter, and run under her Topsails, and Gib, finding that we came up very slow, and were receiving her shot without being able to return them with effect, I ordered the Maintop-Gallant sail set to run up alongside of her.

Isaac Hull's diagram of the engagement between *Constitution* and *Guerriere*, accompanied by a descriptive key. *Constitution* is depicted approaching from windward in the upper right of this drawing. *Reproduced courtesy of the Hull P. Fulweiler Collection at the USS Constitution Museum, Boston, MA*

At 5 minutes past 6 PM being alongside, and within less than Pistol Shot, we commenced a very heavy fire from all of our Guns, loaded with round, and grape, which done great Execution, so much so that in less than fifteen minutes from the time, we got alongside, his Mizen Mast went by the board, and his Main Yard in the Slings, and the Hull, and Sails much injured, which made it difficult for them to manage her. At this time the *Constitution* had received but little damage, and having more sail set than the Enemy she Shot ahead, on seeing this I determined to put the Helm to Port, and oblige him to do the same, or suffer himself to be raked, by our getting across his Bows, on our Helm being put to Port the Ship came too, and gave us an opportunity of pouring in upon his Larboard Bow several Broadsides, which made great havock amongst his men on the forecastle and did great injury to his forerigging, and sails, The Enemy put his helm to Port, at the time we did, but his MizenMast being over the Quarter, prevented her coming too, which

Diagram of the Action between the U.S.S. Constitution & H.B.M.S. Guerriere - 19 Augt. 1812

(B.) Fig. 1st The Constitution, running before the wind with all sail set, sees the Guerriere on a wind, under her topsails, standing to the S. & W.

" Fig. 2 - The Constitution hauls to, shortens sail & prepares for action.

" Fig. 3 - The Constitution commences bearing down upon the Guerriere, who is lying with her main topsail aback, & occasionally waring as in the diagram. The Guerriere commences firing on the Constitution at fig. 3. being about two miles distant.

" Fig. 4 - } " Fig. 5 - } " Fig. 6 - } The Constitution still pressing down upon the Guerriere, and receiving her fire as she wears.

" Fig. 7 - The Constitution along side the Guerriere, first opens her fire and shoots away her mizen mast.

" Fig. 8 - The Constitution along side the Guerriere & to windward, close fighting.

" Fig. 9 - The Constitution, in endeavouring to lay her aboard on the Ld. bow, shoots ahead & crosses her bow. immediately after her fore & main masts fall by the board.

(R.) Fig. 1 - The Guerriere, under double reefs, standing on a wind to the S. & W.

" Fig. 2 - Commences firing on the Constitution, then wears & lays with her M. topsail aback

" Fig. 3 - Fires, & again wears. (as short round as possible).

" Fig. 4 - Bears up before the wind, to make a running fight.

" Fig. 5 - Along side the Constn. - looses her mizen mast.

" Fig. 6 - Constitution attempts to lay her aboard on the Ld. bow but shoots ahead & crosses her bow - immediately after her fore & main masts fall.

brought us across his Bows, with his Bowsprit over our Stern. At this moment I determined to board him, but the instant the Boarders were called, for that purpose, his Foremast, and Mainmast went by the board, and took with them the Gib-boom, and every other Spar except the Bowsprit. On seeing the Enemy totally disabled, and the *Constitution* received but little injury I ordered the Sails filled, to hawl off, and repair our damages and return again to renew the action, not knowing whither the Enemy had struck, or not, we stood off for about half an hour, to repair our Braces, and such other rigging, as had been shot away, and wore around to return to the Enemy, it being now dark, we could not see whether she had any colours flying or not, but could discover that she had raised a small flag Staff or Jurymast forward. I ordered a Boat hoisted out and sent Lieutenant Reed on board as a flag to see whether she had surrendered or not, and if she had to see what assistance she wanted, as I believed she was sinking. Lieutenant Reed returned in about twenty minutes, and brought with him, James Richard Dacres Esqr. Commander of his Britannic Majesty's Frigate the *Guerriere*, which ship had surrendered, to the United States Frigate *Constitution*, our Boats were immediately hoisted out and sent for the Prisoners, and were kept at work bringing them and their Baggage on board, all night. At daylight we found the Enemy's Ship a perfect Wreck, having many Shot holes between wind, and water, and above Six feet of the Plank below the Bends taken out by our round Shot, and her upperworks so shattered to pieces, that I determined to take out the sick and wounded, as fast as possible, and set her on fire, as it would be impossible to get her into Port.

At 3 PM. all the Prisoners being out, Mr. Reed was ordered to set fire to her in the Store Rooms, which he did and in a very short time she blew up. I want words to convey to you the Bravery, and Gallant conduct, of the Officers, and the crew under my command during the action. I can therefore only assure you, that so well directed was the fire of the *Constitution*, and so closely kept up, that in less than thirty minutes, from the time we got alongside of the Enemy (One of their finest Frigates) she was left without a Spar Standing, and the Hull cut to pieces, in such a manner as to make it difficult to keep her above water, and the *Constitution* in a State to be brought into action in two hours. Actions like these speak for themselves which makes it unnecessary for me to say any thing to Establish the Bravery and Gallant conduct of those that were engaged in it, Yet I cannot but make you acquainted with the very great assistance I received from that valuable officer Lieutenant Morris in bringing the Ship into action, and in working her whilst alongside the Enemy, and I am extremely sorry to state that he is badly wounded, being shot through the Body. we have yet hopes of his recovery, when I am sure, he will receive the thanks, and gratitude of his Country, for this and the many Gallant acts he has done in its Service.

Were I to name any particular Officer as having been more useful than the rest, I should do them great Injustice, they all fought bravely, and gave me every possible assistance, that I could wish. I am extremely sorry to state to you the loss of Lieutenant Bush of Marines. he fell at the head of his men in getting ready to board the Enemy. In him our Country has lost a Valuable, and Brave Officer. After the fall of Mr. Bush, Mr. Contee took command of the Marines, and I have pleasure in saying that his conduct was that of a Brave good Officer, and the Marines behaved with great coolness, and courage during the action, and annoyed the Enemy very much whilst she was under our Stern.

Enclosed I have the Honour to forward you a list of Killed, and Wounded, on board the *Constitution*, and a list of Killed, and Wounded, on board the Enemy,[1] with a List of her crew and a Copy of her Quarter Bill, also a report of the damage the *Constitution* received in the Action, I have the honour to be, with very great Respect, Sir, Your Obedient Servant,

Isaac Hull

National Archives and Records Administration, Washington, DC, Record Group 45, Captains' Letters (Microfilm 125, Reel 24, Letter No. 207). In giving *Constitution*'s latitude and longitude on 28 August, Hull's clerk used degree symbols (°) for the figures requiring minute symbols ('). For clarity's sake, this error has been silently corrected.

1. *Niles' Weekly Register* published lists of the killed and wounded on 12 September 1812. *Constitution*'s casualties were reported as 7 killed and 7 wounded, *Guerriere*'s as 15 killed, 62 wounded, and 24 missing.

USS *Constitution* Logbook, 19 August 1812

Remarks on Wednesday August 19th 1812

From Mednght to 4 Am light airs from the Southward and Eastward, and thick foggy weather. A[*lexander*]. T. W[*adsworth*].

From 4 to 8 Am hazey wind from the Southward and Eastward, at 8 Am sett the Top Gallant Sails Thermometer in the air 64 in the water 65°. Geo. C Reed

At ¾ past 8 AM took in the Fore and Mizin Top Gallant Sails, and clewed down the Mizin Top Sail, at 10 Am took the Stream Anchor out of the Fore Chains, took the Stock from it, and put them in the Main hold, at ¾ past 11 AM sett the Mizin Topsail, at Meridian fresh breezes and Cloudy weather. B[*eekman*]. V H[*offman*].

Latitude Observd. 41° 42' North

Commences with fresh breezes from the Northward, and westward and Cloudy, at 2 PM, discovered a sail to the Southward, made all Sail in Chace, at 3 PM perceived the Chace to be a Ship with her Starboard Tacks on board, Close hauled by the wind, at ½ past 3 PM Closing fast with the Chace, who appeared to be a Frigate, at ¼ before 5 PM the Frigate lay her Main Topsail to the Mast, Took in our Top Gallant Sails, stay sails, flying Jib, hauled the Courses up, took the 2d. Reef in the Topsails, and sent down the Royal Yards, and got all snug, and ready for Action and beat to Quarters, at which our Crew gave three Cheers, at 5 PM bore more up bringing the Chace to bear rather off the Starboard Bow, she at that time discovering herself to be our Enemy by hoisting three English Ensigns, at 5 minutes after 5 PM, she discharged her starboard broadside at us without effect, her shot falling Short she immediately wore round, and discharged her Larboard Broadside two shot of which hulled us, and the remainder flying over and through our Rigging, we then hoisted our Ensign and a Jack, at the Fore, and Main Top Gallant Mast heads the Enemy still manouvring to rake us firing alternately his Broadsides, we returning his fire with as many of our Bow Guns from the Main Gun deck as we could bring to bear on her, at ¾ past 5 PM the Enemy Finding his attempts to rake us fruitless, bore up with the wind rather on his Larboard quarter, we then sett our Main Top Gallant Sail, and steered down on his Beam in order to bring him to close action, at 5 minutes after 6 PM hauled down the Jib, and lay the Main Top Sail Shivering and opened on him a heavy fire from all our Guns, at 15 minutes after 6 PM the Enemies Mizen Mast fell over the Starboard side, on which our crew gave three cheers, we then fore reaching on him, attempted raking of his Bow, but our braces being shot away and Jib Haulyards, we could not effect it, he immediately attempted raking of our stern, but failed also, getting but one of his Guns to bear on us which he discharged with little or no effect, having [*h*]is Bowsprit entangled in our Mizen Rigging our Marines during that time Keeping up a very brisk and gauling fire on him, from the Tafferall [Taffrail], and our Boarders preparing to board, at which time Lieutenant Charles Morris, and Lieutenant William S. Bush of the Marines fell from off the Tafferell, the former severely wounded, and the latter Killed, our vessel having way on her shot clear of him, when immediately, it being then 30 Minutes after 6 PM his Fore, and Main Masts fell over on the Starboard side, sett Fore and Main Course, and stood to the Eastward, and took one reef in the Topsails, in order to reeve our Braces, and haulyards which had been shot away; during which time the Enemy a Complete wreck under his Spritsail, fired a Gun in token of Submission to Leeward, which we answered as soon as our Topsails were sett, and our braces rove by wearing Ship, and running under his Lee, hauling up our Courses, and laying our Main Topsail to the Mast, and sending a boat with Lieutenant Reed on board of the prize, at ¼ past 7 PM hoisted out all the Boats, to take out the prisoners, Sent the 2d. and 3d. Cutters on board, with the Surgeons Mate to assist in dressing the wounded, and the Sailing Master to the First Cutter with a Ten Inch Hawser to take the prize in tow, at 8 PM the Boat returned bearing Lieutenant Reed in charge of the prize, and bringing with them Captain Dacres, of; formerly his Britannick Majesty's Ship *Guerriere* mounting 49 Carriage Guns, 30 of which were 18 pounders, on his Main Gun deck 14. 32 pounder Carronades on his quarter deck and one howitzer a 12 pound Calibre also: and 2. 32 pounder

Carronades, and 2 twelve pounder long Guns on his Forecastle Manned with [*Blank*] Men including Marines Boys, and Officers, our loss sustained during the action in Killed, and Wounded 14. Seven of which were Killed, among the latter William S. Bush Senior Lieutenant of Marines, and among the latter Lieutenant Charles Morris dangerously, and Mr. Aylwin Sailing Master, slightly; one of the Seamen of the number Killed Robert Brice lost his life through want of precaution in not Sponging the Gun being blown from the Muzzle of the piece, our Standing and running rigging much cut, and one Shot through the Fore Mast, one through the Main Mast, and one through the heel of the Fore Top Gallant Mast, and the Starboard Cross Jack yard arm cut away, as also the Spare Top Sail Yard in the Main chains, and the Band for the slings of the Main Yard broken, our Spanker Boom, and Gaff Broken by the Enemy when foul of our Mizin Rigging,

National Archives and Records Administration, Washington, DC, Record Group 24, Logbooks and Journals of the USS *Constitution*, 1798–1934 (Microfilm 1030, Reel 1). Heading at top of page reads: "The United States Frigate *Constitution* Isaac Hull Esqr. Commandr." Although this event took place on 19 August 1812, the logbook records it in sea time under the dates 19–20 August.

Captain James R. Dacres, RN, to Vice Admiral Herbert Sawyer, RN

Boston, 7th. September 1812

Sir

I am sorry to inform you of the Capture of His Majesty's late Ship *Guerriere*, by the American Frigate *Constitution*, after a severe action, on the 19th. of August, in Latitude 40°.20' N and Longitude 55.00 West. At 2 PM. being by the Wind on the starboard Tack, we saw a Sail on our Weather Beam, bearing down on us. At 3, made her out to be a Man of War, beat to quarters and prepar'd for Action. At 4, she closing fast, wore to prevent her raking us. At 4.10 hoisted our Colours and fir'd several shot at her. At 4.20 She hoisted her Colours and return'd our fire— Wore several times, to avoid being raked, exchanging broadsides. At 5 She clos'd on our Starboard Beam, both keeping up a heavy fire and steering free, his intention being evidently to cross our bow. At 5.20, our Mizen Mast went over the starboard quarter and brought the Ship up in the Wind.— the Enemy then plac'd himself on our larboard Bow, raking us; a few only of our bow Guns bearing, and his Grape and Riflemen sweeping our Deck. At 5.40, the Ship not answering her helm, he attempted to lay us on board; at this time Mr. Grant who commanded the Forecastle was carried below badly wounded. I immediately order'd the Marines and Boarders from the Main Deck; the Master was at this time shot thro' the Knee, and I receiv'd a severe wound in the back. Lieutenant Kent was leading on the Boarders, when the Ship coming too, we brought some of our bow Guns to bear on her and had got clear of our opponent when at 6.20 our Fore and Main Masts went over the side, leaving the Ship a perfect unmanageable Wreck.— the Frigate shooting a head I was in hopes to clear the Wreck and get the Ship under Command to renew the Action, but just as we had clear'd the Wreck, our Spritsail yard went and the Enemy having rove new Braces &c, wore round within Pistol shot, to rake us,— The Ship laying in the trough of the Sea and rolling her Main Deck Guns under Water and all attempts to get her before the Wind being fruitless; when calling my few remaining officers together they were all of opinion that any further resistance would only be a needless waste of lives, I order'd, though reluctantly, the Colours to be struck.

The loss of the Ship is to be ascrib'd to the early fall of the Mizen Mast which enabled our opponent to choose his position. I am sorry to say we suffer'd severely in kill'd and wounded and mostly whilst she lay on our Bow, from her Grape and Musketry, in all 15 kill'd and 63 wounded, many of them severely, none of the wounded Officers quitted the Deck till the firing ceas'd.

The Frigate prov'd to be the United States Ship *Constitution* of thirty 24 Pounders on her Main Deck, and twenty four 32 Pounders and two 18 Pounders on her upper Deck, and 476 Men— her loss in comparison with ours was triffling, about twenty; the first Lieutenant of Marines and eight kill'd and first Lieutenant and Master of the Ship and eleven Men wounded; her lower Masts badly wounded, and stern much shatter'd and very much cut up about the Rigging.

The *Guerriere* was so cut up, that all attempts to get her in would have been useless. As soon as the wounded were got out of her, they set her on fire, and I feel it my duty to state that the conduct of Captain Hull and his Officers to our Men has been that of a brave Enemy; the greatest care being taken to prevent our Men losing the smallest trifle, and the greatest attention being paid to the wounded who through the attention and skill of Mr. Irvine, Surgeon, I hope will do well.

I hope, though success has not crown'd our efforts, you will not think it presumptuous in me to say the greatest credit is due to the Officers and Ships Company for their exertions, particularly when expos'd to the heavy raking fire of the Enemy; I feel particularly oblig'd for the exertions of Lieutenant Kent, who though wounded early, by a Splinter, continued to assist me in the second Lieutenant, the Service has suffer'd a severe loss; Mr. Scott, the Master, though wounded was particularly attentive and used every exertion in clearing the Wreck as did the Warrant Officers. Lieutenant

Nicoll of the Royal Marines and his party supported the honorable Character of their Corps and the[*y*] suffer'd severely I must particularly recommend Mr. Snow Masters Mate who commanded the foremost Main Deck guns in the absence of Lieutenant Pullman, and the whole after the fall of Lieutenant Ready, to your protection, he having serv'd his time and receiv'd a severe contusion from a Splinter. I must point out Mr. Garby acting Purser to your notice, who volunteer'd his Services on Deck and commanded the after quarter Deck Guns and was particularly active, as well as Mr. Bannister, Midshipman who has pass'd.

I hope in considering the circumstances you will think the Ship entrusted to my charge was properly defended— the unfortunate loss of our Masts, the absence of the third lieutenant, second Lieutenant of Marines, three Midshipmen and twenty four Men considerably weaken'd our Crew and we only muster'd at Quarters 244 Men and 19 Boys on coming into Action; the Enemy had such an advantage from his Marines and Riflemen when close, and his superior sailing enabled him to choose his distance.

I enclose herewith a List of kill'd and wounded on board the *Guerriere*[1] and have the Honor to be, Sir, Your most obedient, humble Servant

(sign'd) Jas. R. Dacres

Public Record Office, London, Adm. 1/502.

1. The original of this enclosure was not found. The names of *Guerriere*'s killed, wounded, and missing were published in *Niles' Weekly Register* on 12 September 1812.

Captain William Bainbridge to Secretary of the Navy Paul Hamilton

U.S. Frigate *Constitution*
St. Salvadore 3d. January 1813

Sir,

I have the honour to inform you that on the 29th. Ultmo. at 2 P,M, in South Latde. 13°.06' & West Longe. 38° about 10 Leagues distance from the coast of Brazils, I fell in with and captured His B.M. Frigate *Java* of 49 Guns, and upwards of 400 men, commanded by Captain Lambert, a very distinguished Officer.— The Action lasted 1h. 55ms. in which time the Enemy was completely dismasted, not having a Spar of any Kind standing.— The Loss on board the *Constitution* was 9 Kill'd and 25 Wounded as per enclosed List.—[1] The Enemy had 60 Kill'd and 101 Wounded certainly (among the latter Captn. Lambert mortally.) but by the enclosed Letter written on board this Ship (by one of the Officers of the *Java*) and accidentally found, it is evident that the Enemies Wounded must have been much greater than as above stated, and who must have died of their wounds previously to their being removed; The Letter States 60 Kill'd and 170 Wounded.—

For further details of the Action, I beg leave to refer you to the enclosed Extracts from my Journal.—[2] The *Java* had in addition to her own crew upwards of 100 Supernumerary Officers & Seamen to join the British Ships of War in the East Indies, also Lieut. General Hislop, appointed to the command of Bombay, Major Walker & Captn. Wood of his Staff, and Captain Marshall, Master & Commr. in the British Navy going to the East Indies to take command of a Sloop of War there.—

Should I attempt to do justice, by representation, to the Brave and good conduct of all my officers & Crew, during the Action, I should fail in the attempt; therefore suffice it to say, that the whole of their conduct was such as to merit my highest encomiums I beg leave to recommend the Officers particularly to the Notice of Government as also the unfortunate Seamen who were wounded, and the family of those Brave Men who fell in the Action.—

The great distance from our own Coast and the perfect Wreck we made the Enemies Frigate, forbid every idea of attempting to take her to the United States, and not considering it prudent to trust her into a Port of Brazils, particularly St. Salvadore as you will perceive by the enclosed Papers No. 1. 2 & 3, I had no alternative but burning her, which I did on the 31st. Ulto. after receiving all the Prisoners and their Baggage, which was very tedious work, only having one Boat left (out of 8) and not one left on board the *Java*—

On Blowing up the Frigate *Java,* I proceeded to this place, where I have landed all the Prisoners on their Parole, to return to England and there remain until regularly exchanged, and not to serve in their professional Capacities in any place or in any manner whatever against the United States of America, until said Exchange is effected— I have the Honor to be, Sir With the Greatest Respect Your Obedt. Hble Svt.

Wm. Bainbridge

P.S. At the time of the Action with the *Java,* I had been seperated 4 days from the *Hornet,* which Vessel at that moment was not within 50 miles of us, but was well occupied in Capturing an English Schooner with a very valuable Cargo, and recapturing an American Ship loaded with Salt; both of them were under convoy of the *Java.*—

National Archives and Records Administration, Washington, DC, Record Group 45, Captains' Letters (Microfilm 125, Reel 26, Letter No. 5). The position of secretary of the navy was vacant at the time Bainbridge penned this letter, Paul Hamilton having resigned on 31 December 1812. William Jones succeeded to the Navy secretaryship on 19 January 1813.

1. Surgeon Amos Evans lists twenty-seven Constitutions wounded in the engagement with *Java*. The two names included on Evans's list but not on Bainbridge's are Midshipman Lewis German and Marine Private Michael Chesley. Amos A. Evans, *Journal Kept on board the United States Frigate "Constitution," 1812* (n.d.; repr., n.p.: Paul Clayton, 1928), 73–74.

2. See the document that follows.

Journal of Commodore William Bainbridge

Extracts from Commodore W. Bainbridges Journal Kept on board the U.S. Frigate *Constitution* [*29–31 December 1812*]

Tuesday 29th. December 1812.— At 9 AM, discovered two strange Sails on the Weather Bow.— At 10 discovered the strange Sails to be ships, one of them stood in for the Land, and the other stood off shore in a direction towards us.— at 10.45 We Tacked Ship to the Nd. & Wd., and stood for the Sail standing towards us.— At 11 AM, Tacked to the Sd. & Ed., hauled up the Mainsail and took in the Royals:— At 11.30 AM, made the Private Signal for the day, which was not answered, and then set the Mainsail and Royals to draw the Strange Sail from the Neutral Coast, and seperate her from the Sail in Company— Wednesday 30th. December 1812 (Nautical time.) In Late. 13°.06' South and Longe. 38° West, 10 Leagues from the Coast of Brazils).— Commences with clear weather and Moderate Breezes from ENE; hoisted our Ensign and Pendant.—[1] At 15 Minutes past Meridian, the Ship hoisted her Colours, an English Ensign, having a Signal flying at her Main[2]

At 1.26 P,M, being sufficiently from the Land, and finding the ship to be an English Frigate, took in the Mainsail & Royals, Tacked ship and stood for the Enemy.— At 1.50 PM, the Enemy bore down with an intention of Raking us, which we avoided by Wearing.— At 2 PM, the Enemy being within half a mile of us and to windward, and having hauled down his Colours (except an Union Jack at the Mizen Mast head), induced me to give orders to the Officer of the 3rd. Division to fire one Gun ahead of the Enemy to make him shew his Colours, which being done brought on a Fire from us of the whole Broadside, on which the Enemy hoisted his Colours and immediately returned our Fire.— A General Action with Round & Grape then Commenced, the Enemy Keeping at a much greater distance than I wished, but could not bring him to closer Action without exposing ourselves to several Rakes.— Considerable manoevres were made by both Vessels to Rake and avoid being Raked.— The following minutes were taken during the Action.—

At 2.10 PM,	Commenced the Action within good Grape & Canister distance, the Enemy to Windward (but much further than I wished.)
At 2.30 "	Our Wheel was shot entirely away.
" 2.40 "	Determined to close with the Enemy, notwithstanding his Raking, set the Fore & Mainsail and Luff'd up close to him.—
At 2.50 PM,	The Enemies Jib Boom got foul our Mizen Rigging
" 3.00 "	The Head of the Enemies Bowsprit & Jib Boom shot away by us
" 3.05 "	Shot away the Enemies foremast by the Board—
" 3.15 "	Shot away his Main Topmast just above the Cap.—
" 3.40 "	Shot away Gafft and Spanker Boom—
" 3.55 "	Shot away his Mizen Mast nearly by the board.—
" 4.05 "	Having silenced the Fire of the Enemy completely, and his Colours in Main Rigging being down, supposed he had struck; then hauled aboard the Courses to shoot ahead, to repair our Rigging which was extremely cut, leaving the Enemy a Complete Wreck; soon after discovered that the Enemies Flag was still flying— Hove too to repair some of our damage.—
" 4.20 "	The Enemies Mainmast went nearly by the Board
" 4.50 "	[*Wore Ship and Stood for the Enemy*—.
" *5.25* "][3]	Got very close to the Enemy in a very effectual Raking Position, athwart his Bows, and was at the very instance of Raking him, when he most prudently Struck his Flag; for had he suffered the Broadside to have Raked him, his additional Loss must have been extremely great, as he laid an unmanageable Wreck upon the Water.—

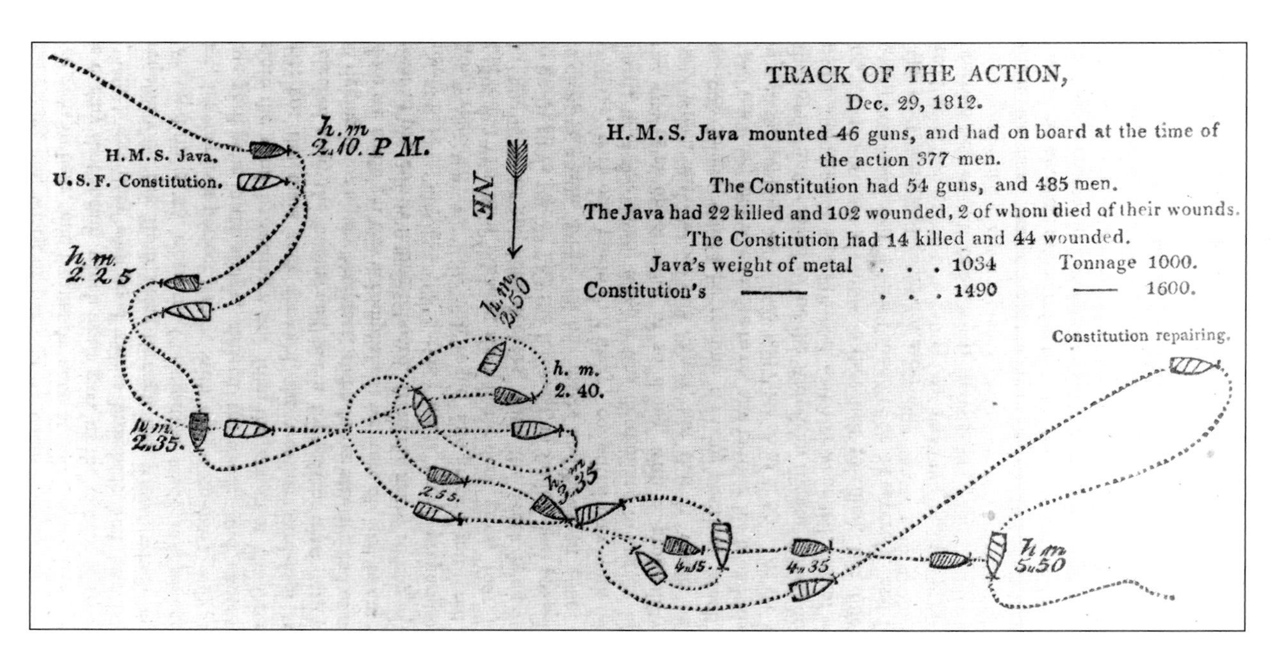

The Naval Chronicle, Vol. 29

After the Enemy had Struck, Wore Ship and Reef'd the Topsails, then hoisted out one of the only two remaining Boats we had left out of 8, and sent Lieut. Parker 1st. of the *Constitn.* to take possession of the Enemy, which proved to be His BM. Frigate *Java,* rated 38 but carried 49 Guns, and mann'd with upwards of 400 men, Commanded by Captain Lambert, a very distinguished Officer, who was mortally wounded.— The Action continued from the commencement to the end of the Fire 1h.55'— The *Constitution* had 9 Kill'd and 25 wounded—[4] The Enemy had 60 Kill'd and 101 certainly Wounded, but by a Letter written on board the *Constitution* by one of the Officers of the *Java,* and accidentally found, It is evident the Enemies Wounded must have been considerably greater than as above stated, and who must have died of their Wounds previously to being removed— The Letter states 60 Kill'd and 170 Wounded.—

The *Java* had her own Compliment of men complete, and upwards of 100 Supernumeraries, going to join the British Ships of War in the East Indies, also several Officers Passengers, going out on Promotion.— The Force of the Enemy in Number of Men at the commencement of the Action, is no doubt considerably greater than we have been able to ascertain, which is upwards of 400 men,— The Officers were extremely cautious in discovering the number. By her Quarter Bill, she had one man more stationed to each Gun than we had.—

The *Constitution* was very much cut in her Sails and Rigging, and many of her Spars injured.— At 7 PM, the Boat returned with Lieut. Chads, the 1st. Lieut. of the Enemies Frigate, and Lieut. General Hislop (appointed Governor of Bombay), Major Walker and Captn. Wood of his Staff.—

Captain Lambert of the *Java* was too dangerously wounded to be removed immediately— The Cutter returned on board the Prize for the Prisoners, and brought Captn. Marshall, Master and Commander of the British Navy, who was Passenger on board as also several other Naval Officers, destine'd for Ships in the East Indies.—

The *Java* was an important ship, fitted out in the completest manner, to carry Lieut. General Hislop and Staff to Bombay; and several Naval Officers for Ships in the East Indies, and had Dispatches for St. Helena, Cape of Good Hope, and every British Establishment in the India and China Seas.—

She had on board Copper for a 74 and two Brigs building at Bombay, and I expect a great many other Valuables; but every thing was blown up in her, except the Officers Baggage, when we set her on fire at 3 P.M, on the 1st. of January 1813. (Nautical time)[5]

National Archives and Records Administration, Washington, DC, Record Group 45, Area File 4 (Microfilm 625, Reel 4, frames 0357–59). There are three extant copies of this document in Record Group 45. One is bound in Captains' Letters, the other two in Area File 4. This transcription was made using the most legible of the three.

1. When converted from sea to civil time, the actual time and date here is noon, 29 December 1812.

2. A representation of *Java*'s signal flag, like the one below, appears

in the text at this point.

Red
Yellow
Red

3. The bracketed text was supplied from the other two copies of this document in Record Group 45.

4. Surgeon Amos Evans lists twenty-seven Constitutions wounded in the engagement with *Java*. The two names included on Evans's list but not on Bainbridge's are Midshipman Lewis German and Marine Private Michael Chesley. Amos A. Evans, *Journal Kept on board the United States Frigate "Constitution," 1812* (n.d.; repr., n.p.: Paul Clayton, 1928), 73–74.

5. When converted from sea to civil time, the actual time and date here is 3 p.m., 31 December 1812.

Lieutenant Henry D. Chads, RN, to First Secretary of the Admiralty John W. Croker

Treplicate United States Frigate *Constitution* off St. Salvador Decr. 31st 1812

Sir

It is with deep regret that I write you for the information of the Lords Commissioners of the Admiralty that His Majestys Ship *Java* is no more! after sustaining an action on the 29th Inst. for several hours with the American Frigate *Constitution* which resulted in the Capture and ultimate destruction of His Majestys Ship. Captain Lambert being dangerously wounded in the height of the Action, the melancholy task of writing the detail devolves on me.

On the morning of the 29th inst. at 8 AM off St. Salvador (Coast of Brazil) the Wind at NE. we perceived a strange sail, made all sail in chace and soon made her out to be a large Frigate; at noon prepared for action the chace not answering our private Signals and tacking towards us under easy sail; when about four miles distant she made a signal and immediately tacked and made all sail away upon the wind, we soon found we had the advantage of her in sailing and came up with her fast when she hoisted American Colours. she then bore about three Points on our lee bow at 1:50 PM the Enemy shortened Sail upon which we bore down upon her, at 2:10 when about half a mile distant she opened her fire giving us her larboard broad-side which was not returned till we were close on her weather bow; both Ships now manauvered to obtain advantageous positions; our opponent evidently avoiding close action and firing high to disable our masts in which he succeeded too well having shot sway the head of our bowsprit with the Jibboom and our running rigging so much cut as to prevent our preserving the Weather gage At 3:5 finding the Enemys raking fire extreemly heavy Captain Lambert ordered the Ship to be laid on board, in which we should have succeeded had not our foremast been shot away at this moment, the remains of our bowsprit passing over his taffrail, shortly after this the main topmast went leaving the Ship totally unmanageable with most of our Starboard Guns rendered useless from the wreck laying over them At 3:30 our Gallant Captain received a dangerous wound in the breast and was carried below, from this time we could not fire more than two or three guns until 4:15 when our Mizen mast was shot away the Ship then fell off a little and brought many of our Starboard Guns to bear, the Enemys rigging was so much cut that he could not now avoid shooting ahead which brought us fairly Broadside and Broadside. Our Main yard now went in the slings both ships continued engaged in this manner till 4:35 we frequently on fire in consequence of the wreck laying on the side engaged. Our opponent now made sail ahead out of Gun shot where he remained an hour repairing his damages leaving us an unmanageable wreck with only the mainmast left, and that toterring; Every exertion was made by us during this interval to place the Ship in a state to renew the action. We succeeded in clearing the wreck of our Masts from our Guns. a Sail was set on the stumps of the Foremast & Bowsprit the Weather half of the Main Yard remaining aloft, the main tack was got forward in the hope of getting the Ship before the Wind, our helm being still perfect. the effort unfortunately proved ineffectual from the Main mast falling over the side from the heavy rolling of the Ship, which nearly covered the whole of our Starboard Guns, We still waited the attack of the Enemy, he now standing towards us for that purpose. on his coming nearly within hail of us & from his manouvre perceiving he intended a position a head where he could rake us without a possibility of our returning a shot. I then consulted the Officers who agreed with myself that on having a great part of our Crew killed & wounded our Bowsprit and three masts gone, several guns useless, we should not be justified in waisting the lives of more of those remaining whom I hope their Lordships & Country will think have bravely defended His Majestys Ship. Under these circumstances. however reluctantly at 5:50 our Colours were lowered from the Stump of the mizen mast and we were taken possession a little after 6. by the American Frigate *Constitution* commanded by Commodore Bainbridge who immediately after ascertaining the state of the Ship resolved on burning her which we had the satisfaction of seeing done as soon as the Wounded were removed. Annexed I send you a return[1] of the killed and wounded and it is with pain I perceive it so numerous also a statement of the comparative force of the two Ships when I hope their Lordships will not think the British Flag tarnished although success has not attended

Force of the two Ships

Java.				Constitution			
Guns		Crew		Guns		Crew	
28 long	18 pours.	Ships Compy.	277	32 long	24 Prs.		
16 Caros.	32 "	Boys	32	22 Caros.	32 "		
2 long	9 "	Supernumery description	68	1 Car	18 "	485	
	46		377	55			
weight of metal	1034			weight of metal	1490		
Tonnage	1000			Tonnage	1490		

This table was appended to Henry Chads's after action report.

us, It would be presumptive in me to speak of Captain Lamberts merit, who, though still in danger from his wound we still entertain the greatest hopes of his being restored to the service & his Country. It is most gratifying to my feelings to notice the general gallantry of every Officer, Seaman & Marine on board. in justice to the Officers I beg leave to mention them individually. I can never speak too highly of the able exertions of Lieuts. Herringham & Buchanan and also Mr. Robinson Master who was severly wounded and Lieuts. Mercer and Davis of the Royal Marines the latter of whom was also severly wounded. To Captn. Jno. Marshall RN who was a passenger I am particularly obliged to for his exertions and advice throughout the action. To Lieutt. Aplin who was on the Main Deck and Lieutt. Sanders who commanded on the Forecastle, I also return my thanks. I cannot but notice the good conduct of the Mates, & Midshipmen. many of whom are killed & the greater part wounded. To Mr. T. C. Jones Surgeon and his Assistants every praise is due for their unwearied assiduity in the care of the wounded. Lieutt. General Hislap, Major Walker and Captain Wood of his Staff the latter of whom was severly wounded were solicitous to assist & remain on the Quarter Deck I cannot conclude this letter without expressing my grateful acknowledgement thus publicly for the generous treatment Captain Lambert and his Officers have experienced from our Gallant Enemy Commodore Bainbridge and his Officers. I have the honor to be Sir Your very obedient Servant

W [*H*] D Chads. 1st. Lieut.
of His Majestys late Ship *Java*

PS. The *Constitution* has also suffered severly, both in her rigging and men having her Fore and Mizen masts, main topmast, both main topsailyards Spanker boom, Gaff & trysail mast badly shot, and the greatest part of the standing rigging very much damaged with ten men killed. the Commodore, 5 Lieuts. and 46 men wounded four of whom are since dead.

Public Record Office, London, Adm. 1/5435.

1. The names of *Java*'s killed and wounded were published in *The Naval Chronicle* 29 (January-June 1813): 348–49.

Captain Charles Stewart to Secretary of the Navy

United States Frigate *Constitution*
[*15*] May 1815.[1]

Sir

On the 20th. of Febuary last, the Island of Madera bearing about WSW distant 60 Leagues we fell in with his Britanic Majesties two Ship of war, the *Cyane* and *Levant,* and brought them to action about 6 OC. in the evening, both of which (after a spirited engagement of forty minuets) surrendered to the Ship under my command.

Considering the advantages, derived by the enemy, from a divided and more active force, as also their superiority in the weight and number of their guns, I deem the speedy and decisive result of this action the strongest assureance which can be given to Government, that all under my Command did their duty, and gallantly supported the reputation of American seamen.—

Inclosed you will receive the minuets of the action, and a list of the Killed and wounded onboard this Ship,[2] also Inclosed you will receive for your information, a statement of the actual force of the Enemy and the number Killed and wounded onboard their ships as near as could be ascertained.[3]

I have the honor to remain Verry Respectfuly Sir Your Most Obdt. Servt.

Chs. Stewart

[Enclosure]

Minutes of the Action between the U.S. frigate *Constitution,* and H.M. Ships *Cyane* and *Levant,* on the 20th

February 1815:—

Commences with light breezes from the Ed. and cloudy weather— At 1 discovered a sail two points on the larboard bow— hauled up and made sail in chace— At ¼ past 1 made the sail to be a ship— At ¾ past 1 discovered another sail ahead—made them out a[*t*] 2 p.m. to be both ships, standing close hauled, with their starboard tacks on board. At 4 p.m. the weathermost ship made signals, and bore up for her consort, then about ten miles to leeward,— we bore up after her, and set, lower, topmast, top gallant, and royal studding sails in chace— At ½ past 4 carried away our main royal mast—took in the sails and got another prepared. At 5 pm. commenced firing on the chace from our two larboard bow guns—our shot falling short, ceased firing— At ½ past 5, finding it impossible to prevent their junction, cleared ship for action, then about 4 miles from the 2 ships— At 40 minutes after 5, they passed within hail of each other, and hauled by the wind on the starboard tact, hauled up their courses, and prepared to receive us— At 45 minutes past 5, they made all sail close hauled by the wind, in hopes of getting to windward of us— At 55 minutes past 5, finding themselves disappointed in their object, and we were closing with them fast, they shortened sail, and formed on a line of wind, about ½ half a cables length from each other. At 6 pm having them under command of our battery, hoisted our colours, which was answered by both ships hoisting English Ensigns. At 5 minutes past 6 ranged up on the starboard side of the sternmost ship, about 300 yards distant and commenced the action by broadsides, both ships returning our fire with great spirit for about 15 minutes,— then the fire of the enemy beginning to slacken, and the great column of smoake collected under our lee, induced us to cease our fire to ascertain their positions and conditions.— in about 3 minutes, the smoake clearing away, we found ourselves abreast of the headmost ship, the sternmost ship luffing up for our larboard-quarter.,— we poured a brodside into the headmost ship, and then braced aback our main and mizen Topsails, and backed astern under cover of the smoake, abreast the sternmost ship, when the action was continued with spirit and considerable effect, until 35 minutes past 6, when the enemy's fire again slackened, and we discovered the headmost bearing up,— felled our Topsails—shot ahead, and gave her two stern rakes— we then discovered the sternmost ship wearing also— wore ship immediately after her, and gave her a stern rake, she luffing too on our starboard bows, and giving us her larboard broadside— we ranged up on her larboard quarter, within hail, and was about to give her our starboard broadside, when she struck her colours, fired a lee gun, and yielded. At 50 minutes past 6, took possession of H.M Ship *Cyane,* captain Gordon Falcon, mounting 34 guns. At 8 pm filled away after her consort, which was still in sight to leeward— At ½ past 8 found her standing towards us, with her starboard tacks, close hauled, with top-gallant sails set, and colours flying— at 50 minutes past 8 ranged close along to windward of her, on opposite tacks, and exchanged broadsides—wore immidiately under her stern and raked her with a broadside, she then crowded all sail, and endeavoured to escape by running— hauled on board our tacks, set Spanker and flying jib in chace— At ½ past 9 commenced firing on her from our starboard bow chaser.— gave her several shot, which cut her spars and rigging considerably— At 10 pm finding they could not escape, fired a gun, struck her colours, and yielded— We immediately took possession of H.M. Ship *Levant,* Honorable captain George Douglas, mounting 21 guns. At 1 a.m. [*21 February*] the damages of our rigging was repaired, sails shifted, and the ship in fighting condition.

[Enclosure]

Minutes of the Chace of the US frigate *Constitution* by an English Squadron of 3 ships, from out the harbour of Port Praya, island of St Iaga:— Sunday March 12th. 1815[4] sea. a/c

Commences with fresh breezes and thick foggy weather— At 5 minutes past 12, discovered a large ship through the fog. standing in for Port Praya;— At 8 minutes past 12, discovered two other large ships astern of her, also standing in for the port. From their general appearance, supposed them to be one of the enemy's Squadrons, and from the little respect hitherto, paid by them to Neutral waters, I deemed it most prudent to put to sea. The signal was made to the *Cyane* and *Levant,* to get under weigh— At 12 after meridian, with our Topsails set, we cut our cable and got under weigh, (when the Portuguese opened a fire on us from several of their batteries on shore,) the prize ships following our motions, and stood out of the harbour of Port Praya, close under East Point, passing the enemy's Squadron about gun shot to windward of them,— crossed our top-gallant yards, and set Foresail, Mainsail, Spanker, Flying Jib, and Topgallant Sails,. The enemy seeing us under weigh,— tacked ship, and made all sail in chace of us,. As far as we could judge of their rates, from the thickness of the weather, supposed them two ships of the line, and one frigate. At ½ past meridian cut away the boats towing astern:— 1st. cutter and gig. At 1 pm found our sailing about equal with the ship on our lee quarter, but the frigate luffing up, gaining our wake, and rather dropping astern of us, finding the *Cyane* dropping fast astern, and to leeward, & the frigate gaining on her fast, I found it impossible to save her if she continued on the same course, without having the *Constitution* brought to action by their whole force, I made the signal at 10 minutes past 1 p.m. to her to tack ship, which was complied with. This manouver, I conceived, would detach one of the enemy's ships in pursuit of her, while at the same time, from her position, she would be enabled to reach the anchorage at Port Praya, before the detached ship could come up with her, but if they did not tack after her, it would afford her an opportunity to double their rear, and make her escape before the wind. They all continued in full chace of the *Levant,* and this ship: The ship on our lee quarter, firing by divisions, broadsides—her shot falling short of us. At 3 pm by our having dropped the *Levant* considerably, her situation became (from the position of the enemy's frigate,) similar to the *Cyane,* it became necessary to seperate also from the *Levant,* or risk this ship being brought to action to cover her. I made the signal at 5 minutes past 3 for her to tack, which was complied with. At 12 minutes past 3 the whole of the enemy's squadron tacked in pursuit of the *Levant,* and gave up the pursit of this ship. This sacrifice of the *Levant* became necessary for the preservation of the *Constitution.* Sailing Master Hixon, Midshipman Varnum, 1 Boatswains mate, and 12 men, were absent on duty in the 5th. cutter to bring the cartel brig under our stern.—

National Archives and Records Administration, Washington, DC, Record Group 45, Captains' Letters (Microfilm 125, Reel 44, Letter No. 93 and enclosures). When Charles Stewart sailed from Boston in December 1814, the Navy secretaryship was vacant, William Jones having left office earlier that month. Stewart addressed his letter, therefore, to "the Honbl. Sectry. of the Navy." Jones's replacement as Navy secretary was Benjamin W. Crowninshield who assumed his new duties on 16 January 1815.

1. *Constitution* anchored off Sandy Hook on 15 May 1815. The ship's logbook records that Stewart's dispatch was sent off to New York City via the customs boat that day.

2. Stewart reported *Constitution*'s casualties as 3 killed and 12 wounded (3 mortally). The ship's logbook records 4 killed and 12 wounded in the engagement. The original of this enclosure was not found. It was published in *Niles' Weekly Register,* 27 May 1815, pp. 218–19.

3. Stewart gave *Levant*'s force as 21 guns and 155 men, and *Cyane*'s force as 34 guns and 180 men. He listed the casualties in *Levant* as 23 killed and 16 wounded, and for *Cyane* 12 killed and 26 wounded. The original of this enclosure was not found. It was published in *Niles' Weekly Register,* 27 May 1815, p. 219.

4. These minutes were dated according to sea time. The actual date for this event is 11 March 1815.

USS *Constitution* Logbook, 20 February 1815

Monday February 20th 1815

First part moderate breezes and cloudy with a little haze. At 1 P.M. a sail in sight to the Sd & Wd. hauled up for her and gave chase, set staysails &c At 1.15. made her out a large ship— 1.30 discovered another ship to the westward of her both standing close hauled towards us under a press of sail with their starboard tacks on board. At 3 the weather ship made signals, at 4 she bore up making signals to the ship to leeward and firing guns;— bore up after her and crouded all sail in chase, set lower, topmast, topgallant, and royal studding sails. At 4.15 the lee ship tacked to the Southward. At 4.30 carried away our Main royal mast. At 5 fired on the chase from the first gun 1st division and the chase gun on the Forecastle, our shot falling short ceased firing; the lee ship tacked to the Northd. At 5.40. they closed, passed within hail of each other, shortened sail, hauled up their courses, and appeared to be making preparations to receive us At 5.45 they set staysails, hauled aboard their tacks and endeavoured to outwind us— 5.55. they shortened sail and formed on a line of wind at half a cables length from each other— At 6 sent up our colours they hoisting at the same time red English Ensigns— At 6.5. ranged within 300 yards upon the starboard side of the sternmost ship and invited the action by firing a shot between the two ships which immediately commenced with an exchange of broadsides— in about 15 minutes their fire became slack, ordered our batteries to cease firing until the smoke cleared away; finding ourselves abreast of the headmost ship gave her our broadside, backed the after yards and closed with the sternmost ship under cover of the smoke;— the action was renewed with additional vivacity on both sides and continued until 6 h. 35 m when their fire again slackened, and we discovered the headmost ship bearing up; filled our after sails, shot ahead, and gave her two broadsides into her stern; the sternmost ship was then discovered to be waring, wore short round after her, she luffed to on our starboard bow and fired her larboard broadside, luffed to on her larboard quarter within 50 yards, when she struck her colours, hoisted a light, fired a lee gun and yielded. At 6 h. 50 m. took possession of His Britannic Majesty's Ship *Cyane* Captain Gordon Falcon mounting 34 carriage guns and two swivels, got out all the officers, put fifteen marines over her prisoners, and gave her in charge to Lieut. Hoffman with a small crew. At 7. 45 m. filled away after her consort and at 8 discovered her with damages repaired and topgallant sails set standing towards us— at 8 h. 40 m passed on opposite tacks within 50 yards to windward of her, exchanged broadsides, were

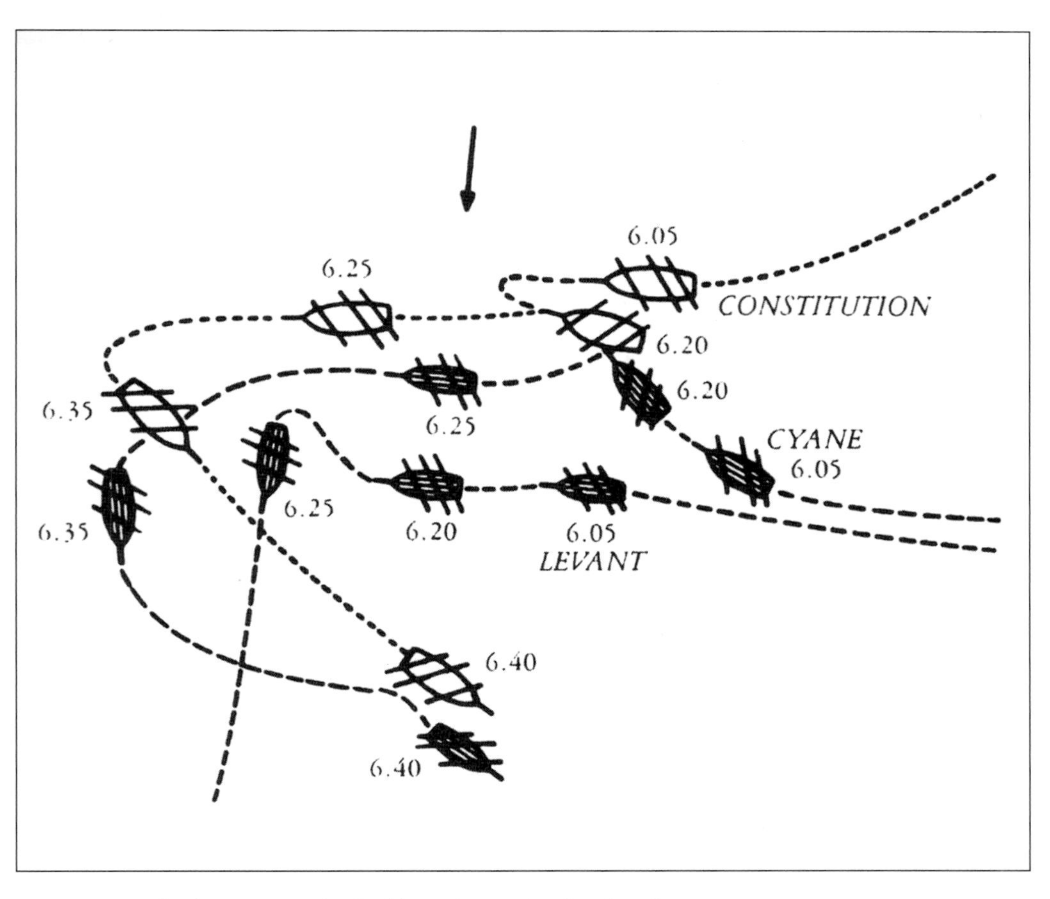

Diagram of *Constitution*'s engagement with HMS *Cyane* and *Levant*, 20 February 1815. *Roosevelt, The Naval War of 1812*

under her stern and raked her, she made all sail and commenced running; set the Courses, Spanker & flying Jib in Chase— at 9. 30. opened a fire upon her from our chase guns:— At 10 finding escape impossible she fired a gun to leeward and yielded;— took possession of His Majesty's Ship *Levant* Honble. George Douglass Captain, mounting 21 carriage guns. At 11 all hands employed repairing damages securing the prisoners &c. &c.

At 1 A.M. [*21 February*] the ship was put in good fighting condition

National Archives and Records Administration, Washington, DC, Record Group 24, Logbooks and Journals of the USS *Constitution*, 1798–1934. (Microfilm 1030, Reel 1). Heading at top of page reads: "Remarks &c on board U.S. frigate *Constitution*, Charles Stewart Esqr. Commander on a Cruise." Although the action took place on 20 February 1815, the logbook records it in sea time under the date of 21 February.

USS *Constitution* Logbook, 11 March 1815

Saturday, March 11th 1815

At 8 A.M. blacked the starboard bends. Sent Captains Douglass & Falcon on shore; at Meridian they returned and said they had arranged for the Cartel. Sent a boat and an Officer to bring the English brig under our stern to have her convenient to provision &c. Made preparations for supplying the Cartels from the prizes.

Commences with fresh breezes and thick foggy weather. At 0 h. 5 m. P.M. discovered a large ship through the fog standing in for Port Praya. At 0 h. 8 m. discovered two other large ships astern of her also standing in for the Port. From their general appearance supposed them to be one of the Enemy's squadrons, and from the little respect hitherto paid by them to neutral waters I deemed it most prudent to put to sea. The signal was immediately made to the *Cyane* and *Levant* to get under weigh. At 0 h. 12 m with our topsails set we cut our cable and got under weigh, when the Portuguese opened a fire upon us from several of their batteries on shore. The prize ships followed our motions and stood out of the harbour of Port Praya close under East point passing the Enemy's squadron about gun shot to windward of them; crossed our topgallent yards and set Foresail, Mainsail, Spanker, Flying-jib and topgallant sails. The Enemy seeing us under way tacked ship and made all sail in chase of us. As far as we could judge of their rate from the thickness of the weather supposed them to be two ships of the line and one frigate. At 0 h. 30 m. cut away the boats towing astern, first cutter and gig. At 1 P.M. we found our sailing about equal with the ship on our lee quarter, but the frigate luffing up and gaining our wake and rather dropping astern of us. The *Cyane* dropping fast astern and to leeward and the frigate gaining on her fast I found it would be impossible to save her if she continued on the same course without having the *Constitution* brought to action by their whole force, I made the signal at 1 h. 10 m to her to tack which was complied with. This manoeuvre I conceived would detach one of the Enemy's ships in pursuit of her, while at the same time from her position she would be enabled to reach the anchorage at Port Praya before the detached ship would come up with her; but if they did not tack after her it would afford her an opportunity to double their rear and make her escape before the wind. They all continued in full chase of the *Levant* and this ship, the ship on our lee quarter firing her broadside by divisions the shot falling short of us. At 3 having dropped the *Levant* considerably her situation became from the position of the Enemy's frigate similar to the *Cyane*'s. It now became necessary to separate also from the *Levant* or risk this ship being brought to action to cover her; the signal was accordingly made at 3 h. 5 m. P.M for her to tack which was complied with. At 3 h. 12 m. the whole of the Enemy's squadron tacked in pursuit of the *Levant* and gave over the pursuit of this ship. This sacrifice of the *Levant* became necessary for the preservation of the *Constitution*. Set the royals and kept Large from the wind. Sailing master Hixon, Midshipman Varnum, one Boatswain's mate and twelve men, who were absent on duty in the 5th cutter to bring the cartel brig under our stern were left on board the *Levant,* which ship they reached before she cut. Surgeon's mate Johnson, with the Sailmaker and his mate, were likewise on board the *Levant*.

Latter part moderate breezes and hazy.

National Archives and Records Administration, Washington, DC, Record Group 24, Logbooks and Journals of the USS *Constitution*, 1798–1934 (Microfilm 1030, Reel 1). Heading at top of page reads: "Remarks on board U.S. frigate *Constitution*, Charles Stewart Esqr. Commander." Although this event took place on 11 March 1815, the logbook records it in sea time under the dates of 11 and 12 March.

Captain George Douglas, RN, to First Secretary of the Admiralty John W. Croker

United States Frigate *Constitution*
at Sea 22nd. February 1815.—

Sir,

It is with extreme regret I have to acquaint you for the information of their Lordships of the Capture of His Majesty's Ships *Levant* and *Cyane* on the night of the 20th. Inst., by the United States Frigate *Constitution*, in Latitude 33° 17' N. and Longitude 13°. 10' W.— His Majesty's Ships sailed in Company from Gibralter Bay where they had been refitting on the 16th. of February and from Tangier Bay on the 17th. with the wind at S.E. shaping a course for Madeira, a Swedish Brig was the only Vessel seen until the afternoon of the 20th. February, the *Cyane* then about ten miles on the weather Beam looking out, at about 1H. 30' she made the Signal for a strange sail N.W. the *Levant* was immediately hauld close upon a wind on the Starboard tack, at 1H. 45' PM. a sail was seen on the weather beam standing apparently on a wind on the Larboard Tack, and closing with the *Cyane*, about 3 PM. observed the *Cyane* bear up, and the stranger bear up after her, about 3H. 15', having brought both Ships abaft the beam the *Levant* was tacked, to close with them, answered the Signal No. 377 from the *Cyane,* and cleared for Action, at 4 PM. tack'd again, and at 4H. 15' spoke His Majesty's Ship *Cyane*, when Captain Falcon informed me he had every reason to believe the stranger was an American Frigate, but owing to the very hazey state of the weather, it was impossible at that time to make out her exact force, the stranger still coming down upon us, and His Majesty's Ships continuing close to each other, and running free under easy sail, with the intention if possible, of forcing a night Action, at 5H.10 the Stranger hoisted American Colours, and appeared to be a Frigate of the largest class, the Colours of His Majesty's Ships being hoisted at the same time, at 5.15 PM. the action commenced with the three Ships, the *Levant* taking a position upon the Enemy's larboard bow, and the *Cyane* a little abaft his larboard Beam, which was kept up with great spirit on all sides, until about 6.40 PM. when finding that the whole of the runing rigging, and greater part of the standing rigging were shot away, and the Masts and yards considerably injured,— I put the Ship before the wind in order to get her in a governable state, and stop up the shot holes, more effectually, having by this time received several between wind and water, and the Ship leaking considerably, as the smoke cleared away I observed that the *Cyane*, in attempting to get before the wind also, had in consequence of all her runing rigging being shot away, unavoidably come too, on the larboard Tack with all sails aback, and the Ship apparently unmanageable by which means we were unfortunately seperated, the enemy at this time appeared to have suffered but little, in consequence of his being to windward during the action, and Keeping at too great a distance, to allow our Carronades to do full execution. Before it was in my power to haul the *Levant* to the wind, I observed the Enemy range up close to the *Cyane*, and pass her without any guns being fired, but it being night I could not discover whether the Colours of the *Cyane* were still flying or not, at 8.15 PM the *Levant* being again ready for Action the Ship was hauled to the wind, and tack'd to close with the Enemy, at that time standing towards us, and at 9.10 PM finding it was out of my power to weather him, pass'd close under his lee and gave him our Starboard Broadsides, as long as the guns would bear, receiving at the same time a most heavy and destructive fire from the Enemy both in the rigging and Hull, at 9.30 PM finding that the *Cyane* had undoubtedly been obliged to strike her colours, the *Levant* was again put before the wind with the hopes of saving the Ship, receiving several heavy raking Broadsides in wearing from the enemy, who were in chace of us, every effort was now made to make all sail, but owing to the crippled state the Ship was again in, the whole of the lower and runing rigging, the Wheel, Main topgallant yard, Mizen Topmast and Starboard foretopmast Studdingsail boom being shot away, the lower Masts much wounded and the sail shot and torn to peices, caused an unavoidable delay, the enemy keeping up a constant fire with his bow Guns, during the chace, and coming fast up with us, at 10.20 PM. seeing that the enemy was ranging up on our larboard quarter with the intention of giving us his broadside, and having consulted the opinion of the Officers who agreed with me that any further resistance would only be an useless sacrafice of more lives, at 10.40 the Colours were hauld down and the Ship taken possession of by the United States Frigate *Constitution*, mounting 52 Guns, and a Complement of 472 Men,— Although I was aware of the superiority of the Enemy's force, I nevertheless conceived it my duty to bring him to action, with the hopes of at least disabling him and preventing his intercepting, two valuable Convoys which sailed from Gibralter on the same day with the *Levant*, which I knew to be in our neighbourhood, this object was fortunately accomplished,— Although it was my misfortune to be obliged to strike my Colours to the *Constitution*, I cannot omit mentioning in the highest terms Lieutenants John Henderson 2nd. and Wm. Jones, acting, also Mr. James Stannes, Acting Master and Lieutenant Wm. Meheuse, Royal Marines, and the Petty Officers, Seamen, and Marines of His Majesty's late Ship *Levant* for their very gallant behaviour, during a most unequal contest, and constant fatigue of five hours and a half, likewise to Captain Falcon of His Majesty's Ship *Cyane* for the very able support I received from that Ship during the action until the unfortunate, but unavoidable seperation of the two Ships, a copy of whose letter I enclose, from that period, until the surrender of the *Cyane*,—[1] I likewise enclose a list of the names of the killed and wounded on board the two Ships.—[2] The *Levant* at the Commencement of the action being eleven short of Complement, and the *Cyane* thirteen.—* I have the Honor to remain, Sir, Your most Obedient Humble Servant.

George Douglas Captain

*NB. It has never been my power to ascertain the exact loss of the *Constitution*, but as far as I could learn she had from four to six killed and about twelve Wounded, three of whom died during the time I was on board the Ship.— The *Constitution* likewise suffered considerably in her rigging, and sails, with a number of shot in her hull, which it is to be regretted owing to the distance, had not the desired effect, several struck her between wind and water in consequence of which the pumps were frequently at work.

Public Record Office, London, Adm. 1/1740.

1. See the following letter, Falcon to Douglas, 22 February 1815.

2. Douglas listed *Levant*'s casualties as 6 killed and 18 wounded. Falcon listed *Cyane*'s casualties as 4 killed and 20 wounded.

Captain Gordon T. Falcon, RN to Captain George Douglas, RN

United States Ship of War
Constitution 22nd. February 1815.

Sir

As the distance of the *Levant* on the night of 20th. inst. must have prevented your seeing the state and situation of the *Cyane* or knowing the circumstances that led to her capture, I take the earliest opportunity of laying before you the particulars of that unfortunate event.—

During the greater part of the action the situation of the two Ships was such as to afford you an opportunity of perceiving every occurence, it therefore appears unnecessary for me to say more, than that it was my constant endeavour to close with the Enemy, finding we were too far distant for Carronades, at the same time exposed to the full effect of his long guns, & obtain a position on his quarter, in this however I was only partially successful, as the situation and superior sailing of the enemy's Ship enabled him to Keep the *Cyane* generally on his broadside, consequently exposed to a heavy fire, from which in the early part of the action the Ship suffered very much in the rigging and latterly in the hull.—

At about 40 minutes after 6 I was informed the *Levant* had bore up, and upon observing her situation as well as I could through the smoke &c., I imagined from what I could discover, it was your intention to ware, in consequence I immediately did so, in performing which I had the mortification to find that not a brace or a bowline, except the larboard fore brace, were left, but observing the *Levant* was exposed to a heavy raking fire, the *Cyane* was brought to the wind on the larboard tack, unfortunately with all the sails aback, with the intention of covering the former, in which situation the action was maintained so long as a gun would bear, the smoke under the lee preventing my discovering for some time that the *Levant* was continuing before the wind, on seeing which I endeavoured to follow her, but owing to the situation of the sails and crippled state of the rigging it was not in my power to get the Ship before the wind,— A short cessation of firing having taken place I embraced the opportunity of reeving fresh braces &c. in the hope of getting the Ship again under Command, but before this could be accomplished the enemy having wore, again opened his fire on the *Cyane,* and closing, took a position within hail on the larboard quarter, which it was impossible to prevent, and equally so under such circumstances, to refit the Ship, having nearly the whole of the standing and all the runing rigging cut, the sails very much shot and torn,— all the lower Masts severely wounded, particularly the Main and Mizen masts, both of which were tottering Fore yard, Fore and Mizen topmasts, Gaff and Driver boom, Main topgallant yard and fore topgallant mast, shot away or severely wounded,— a number of shot in the hull eight or nine between wind and water,— six guns dismounted or otherwise disabled by shot, drawing of bolts &c.— with a considerable reduction from our strength in killed and wounded,— In this state the *Cyane* was when the Enemy's Ship took the position already mentioned,— the *Levant* nearly two miles to leeward and still going before the wind, therefore not in a situation to afford support, without which I could have no reasonable prospect of making any impression on so very superior a force as I was then singly opposed to, even had the *Cyane* been perfectly effective, which to all appearance the Enemy still was;— Thus situated it was with much concern I foresaw the surrender of His Majesty's Ship as an event to which I should be obliged to submit— not relying however on my own judgment I consulted my Officers and finding they were of opinion that the situation and crippled state of the Ship prevented any prospect of success against a force considerably more than double our own, or even of effecting our escape from a Ship so much superior in sailing— conceiving under such circumstances that farther resistance would be attended only with a loss of lives equally unavailing and unnecessary—and feeling confident that every thing in my power had been attempted against the Enemy though without the desired effect—and imagining the *Levant* to be at such a distance as to insure her escape should you consider such a step proper, I, on a due consideration of these circumstances, felt it my painful duty to direct the colours to be struck, which was done at 7' OClock, and the Ship soon after taken possession of by the United States Ship of War *Constitution.*—

Having thus endeavoured to give you a particular statement of the circumstances relating to the unfortunate capture of His Majesty's late Ship, under my Command, I proceed to a more pleasing duty, that of making known to you the Bravery, and good conduct of the Commissioned, Warrant, and Petty Officers, Ships Company and Royal Marines under my Command, and of assuring you that during this unequal contest their exertions were such as merit my warmest approbation, affording me the strongest assurance that had the *Cyane* been fortunate enough to have met a more equal force, or even on the present occasion, along with the *Levant,* to have succeeded in bringing the Enemy to closer action, the result would have been very different;— From Lieut. Alexr. McKenzie 1st. Lieut. who was slightly wounded, I received every assistance, and beg in the strongest manner to recommend him to the notice and protection of their Lordships;— I feel at the same time much pleasure in naming Lieut. Henry Jellicoe 2nd. Lieut., Mr. George Smith, acting Master, and Lieut. Peter Meares, Royal Marines, each of whom in their respective situations conducted themselves much to their own credit and my satisfaction,— I likewise take this opportunity of bringing to their Lordships notice Messrs. John Lingard and J. W. H. Handley, Masters Mates and Mr. Joseph Walker, Midshipmen, they having all passed for Lieutenants, and deserving of every credit, the two first are severely wounded, particularly Mr. John Lingard, who has served nearly five years with myself and whose conduct upon all former occasions, as well as the present, entitles him to my warmest recommendation, as a most promising young Officer.

I have the honor to enclose a return of the killed and wounded[1] which it is a satisfaction to remark is short of what might have been expected, some of the latter are severe I am happy however to be able to state that they are under the care of Mr. [Tegetmeir], Surgeon of the *Cyane,* all doing well.— The *Cyane* went into action thirteen short of Complement and four unable to attend at their quarters from sickness.

Trusting that the above statement may prove satisfactory, and confidently hoping that the *Cyane*s conduct in the action may be entitled to your approbation, and favourable report.— I have the honor to remain Sir, Your most Obedient Servant.

Gordon Falcon <u>Captain</u>

Public Record Office, London, Adm. 1/1740.

1. Falcon listed *Cyane*'s casualties as 4 killed and 20 wounded.

Remarks on Board HMS *Cyane*, 20 February 1815

Remarks, &ca. on board the *Cyane,*
Monday 20th. February 1815.

PM. moderate Breezes from E.N.E and hazy weather, all sail set steering WNW at 1, exercised great Guns and small arms;— 1.20 a strange sail to the Northward—made compas signal NW to the *Levant.* at the time hull down to WSW in studding sails and

hauled to wind[*ward*] on the starboard tack in chace, the stranger appearing to be a square rigged vessel steering to the So.ward with a fore royal set, but no main the weather continued hazy, and could at times see the chace only very indistinctly;— ship head from North to NNW. moderate Breezes from ENE to NE. with some swell:— at ¼ p[ast] 2 made the chace out to be a ship and shortly after a ship of war, and at about ½ past 2 to be a Frigate: made the private signal which not being answered, at ¾ past 2, bore up to close the *Levant,* then on a wind on the starboard tack to leeward hull down, the stranger bearing NE. by N distance 5. or 6 miles.— observed her to shape a course after us, and to make all sail,— cleared Ship for action— made Signals No. 3. 11. and 377 to the *Levant* with Guns; observed a flag at the main on board the *Levant* which could not on account of the haze be distinguished. soon after bearing up, the Stranger hoisted the flags white pierced red, over half red half blue at the fore;— found she was gaining fast upon us, made all possible sail; observed him to carry away his main royal mast, shortly after he fired his chace Guns, the shot falling far distant of the ship;— trimmed sails as necessary, making the most direct course to close the *Levant,* then working to windward made all sail: at ¾ past 4 shortened sail to topsails, top-gallant sails and foresails; spoke the *Levant* and informed Captain Douglas that we had every reason to suppose the ship to windward was an American frigate, but that I had not been able to ascertain her force as she had kept nearly always end on to us; when finding it was his intention to engage her, the Ships Companies cheered; hauled to the wind and set the mainsail keeping about ½ Cables length astern of the *Levant;* the Stranger hauled up a little likewise and set his Mainsail and spanker:— found we could not gain the wind of him, again spoke the *Levant,* when both ships bore up together with the view of prolonging the commencement of the action till night: the Stranger s[t]ill continuing to close us fast, and finding we could not succeed in delaying the action so long as wished for, at 5.10 hauled to the Wind, on the starboard tack and up mainsail:— the stranger hoisting American colours and hauling up likewise discovered to us his force viz. 15 Guns on the main deck, 8 on the quarter deck, and 4 on the forecastle of a side:— hoisted the Colours; at 5.20 the enemy being about ⅓ of a mile, or rather more distant, and a point abaft the beam of the *Cyane* tried the range of his shot, and immediately afterwards commenced action; both Ships returning his fire the *Levant* holding an advantageous position on the bow, the *Cyane* a little abaft the beam of the Enemy's braced sharp up, and endeavoured to close the Enemy's quarter, observing our Shot to frequently to fall short, whilst he held a decided superiority in his long guns; cutting our rigging and sails to pieces:— at about 6 shivered the main-topsail and allowed the Enemy to draw a little ahead, when we again filled and endeavoured to luff upon his quarter, upon discovering which he threw all aback, and again brought the *Cyane* on his beam; found we had neared him a little as his Musquet Balls began to fall on board, at the same time we were suffering very much in the Hull and rigging, his fire being greatly superior, and ours rather slacker in consequence of some of the Guns being disabled:— at about 40 minutes past 6 the *Levant* appear'd to be keeping away, and imagining it might be with the intention of wearing, wore immediately:— found we had not a brace or bowline left, except the larboard fore brace, but the *Levant* being at this time exposed to heavy raking fire, the Enemy having filled across her, the *Cyane* was brought to the Wind on the Larboard tack, with everything aback, for the purpose of covering the *Levant,* renewed the action and continued it so long as the Guns would bear; lost sight for some time of the *Levant* in the smoke;— shortly afterwards the firing ceased for a short time, discovered the Enemy's Ship had wore and was standing for the *Cyane,* and soon after commenced firing her Starboard Guns; turned the hands up to refit rigging, rove new braces, &ca. the Enemy again ceased firing; upon the clearg. up of the smoke found the *Levant* was running to leeward; attempted to get the ship before the Wind to close her, which, owing to the crippled state of the rigging and situation of the sails lying flat aback, and Driver &c entangled in the Wreck of rigging, [&] about the Mizen mast as not to be able to get it down; Jib sheets & flying Jib Hallyards shot away, could not be accomplished, before the Enemy had closed us, and taken a position on our larboard quarter, within hail:— the ship at this time totally unmanageable with most of the standing and all the running rigging shot away, sails much shot and torn, all the lower masts, and several of the yards severely wounded particularly the main and mizen and fore and mizen topmasts (the latter fell soon after) and fore topgallant mast, foreyard, cross Jack and main topgallant yards; spanker boom and gaff:— a number of shot in the Hull, nine or ten between wind and water, five guns disabled by Enemy's shot, drawing of Bolts from the side and starting of Chocks, &ca., starboard clue of fore topsail shot away and topsail sheets:— the Enemy holding a position in which the Ship was exposed to his broadside with not more than 3 or 4 Guns to bear upon him, the *Levant* at this time nearly two miles directly to leeward, and still going before the wind, and the Enemy to all appearance having sustained but little damage, and in full command of the Ship:— Thus situated without an opportunity of refitting the rigging so as to get the Ship under command, further resistance was considered as useless against such a superior force, a light was therefore shewn, and at 7 o'Clock the colours were struck: Shortly afterwards an officer came on board when it was found we were captured by the United States Ship of War *Constitution* Captain Charles Stewart, mounting 52 Guns, long 24 Pounders, and 32 Pounders, Carronades with a Complement at the Commencement of the action of 472 Men: At the time of quitting the *Cyane,* the *Levant* was still going to Leeward.

Signed. Gordon Falcon
Captain.

Public Record Office, London, Adm. 1/5449.

An Act for the Better Government of the Navy of the United States, 23 April 1800.

SECTION 1. *Be it enacted by the Senate and House of Representatives of the United States of America in Congress assembled,* That from and after the first day of June next, the following rules and regulations be adopted and put in force, for the government of the navy of the United States.

Art. I. The commanders of all ships and vessels of war belonging to the navy, are strictly enjoined and required to show in themselves a good example of virtue, honour, patriotism and subordination; and be vigilant in inspecting the conduct of all such as are placed under their command; and to guard against, and suppress, all dissolute and immoral practices, and to correct all such as are guilty of the them, according to the usage of the sea service.

Art. II. The commanders of all ships and vessels in the navy, having chaplains on board, shall take care that divine service be performed in a solemn, orderly, and reverent manner twice a day, and a sermon preached on Sunday, unless bad weather, or other extraordinary accidents prevent it; and that they cause all, or as many of the ship's company as can be spared from duty, to attend at every performance of the worship of Almighty God.

Art. III. Any officer, or other person in the navy, who shall be guilty of oppression, cruelty, fraud, profane swearing, drunkenness, or any other scandalous conduct, tending to the destruction of good morals, shall, if an officer, be cashiered, or suffer such other punishment as a court martial shall adjudge; if a private, shall be put in irons, or flogged, at the discretion of the captain, not exceeding twelve lashes; but if the offence require severer punishment, he shall be tried by a court martial, and suffer such punishment as said court shall inflict.

Art. IV. Every commander or other officer who shall, upon signal for battle, or on the probability of an engagement, neglect to clear his ship for action, or shall not use his utmost exertions to bring his ship to battle, or shall fail to encourage, in his own person, his inferior officers and men to fight courageously, such offender shall suffer death, or such other punishment as a court martial shall adjudge; or any officer neglecting, on sight of any vessel or vessels of an enemy, to clear his ship for action, shall suffer such punishment as a court martial shall adjudge; and if any person in the navy shall treacherously yield, or pusillanimously cry for quarters, he shall suffer death, on conviction thereof, by a general court martial.

Art. V. Every officer or private who shall not properly observe the orders of his commanding officer, or shall not use his utmost exertions to carry them into execution, when ordered to prepare for, join in, or when actually engaged in battle; or shall at such time, basely desert his duty or station, either then, or while in sight of an enemy, or shall induce others to do so, every person so offending shall, on conviction thereof by a general court martial, suffer death or such other punishment as the said court shall adjudge.

Art. VI. Every officer or private who shall through cowardice, negligence, or disaffection in time of action, withdraw from, or keep out of battle, or shall not do his utmost to take or destroy every vessel which it is his duty to encounter, or shall not do his utmost endeavour to afford relief to ships belonging to the United States, every such offender shall, on conviction thereof by a general court martial, suffer death, or such other punishment as the said court shall adjudge.

Art. VII. The commanding officer of every ship or vessel in the navy, who shall capture, or seize upon any vessel as a prize, shall carefully preserve all the papers and writings found on board, and transmit the whole of the originals unmutilated to the judge of the district to which such prize is ordered to proceed, and shall transmit to the navy department, and to the agent appointed to pay the prize money, complete lists of the officers and men entitled to a share of the capture, inserting therein the quality of every person rating, on pain of forfeiting his whole share of the prize money resulting from such capture, and suffering such further punishment as a court martial shall adjudge.

Art. VIII. No person in the navy shall take out of a prize, or vessel seized as a prize, any money, plate, goods, or any part of her rigging, unless it be for the better preservation thereof, or absolutely necessary for the use of any of the vessels of the United States, before the same shall be adjudged lawful prize by a competent court; but the whole, without fraud, concealment, or embezzlement, shall be brought in, and judgment passed thereon, upon pain that every person offending herein shall forfeit his share of the capture, and suffer such further punishment as a court martial, or the court of admiralty in which the prize is adjudged, shall impose.

Art. IX. No person in the navy shall strip of their clothes, or pillage, or in any manner maltreat persons taken on board a prize, on pain of such punishment as a court martial shall adjudge.

Art. X. No person in the navy shall give, hold, or entertain any intercourse or intelligence to or with any enemy or rebel, without leave from the President of the United States, the Secretary of the Navy, the commander in chief of the fleet, or the commander of a squadron; or in case of a vessel acting singly from his commanding officer, on pain of death, or such other punishment as a court martial shall adjudge.

Art. XI. If any letter or message from an enemy or rebel, be conveyed to any officer or private of the navy, and he shall not, within twelve hours, make the same known, having opportunity so to do, to his superior or commanding officer; or if any officer commanding a ship or vessel, being acquainted therewith, shall not, with all convenient speed, reveal the same to the commander in chief of the fleet, commander of a squadron, or other proper officer whose duty it may be to take cognizance thereof, every such offender shall suffer death, or such other punishment as a court martial shall adjudge.

Art. XII. Spies, and all persons who shall come or be found in the capacity of spies, or who shall bring or deliver any seducing letter or message from an enemy or rebel, or endeavour to corrupt any person in the navy to betray his trust, shall suffer death, or such other punishment as a court martial shall adjudge.

Art. XIII. If any person in the navy shall make or attempt to make any mutinous assembly, he shall, on conviction thereof by a court martial, suffer death; and if any person as aforesaid shall utter any seditious or mutinous words, or shall conceal or connive at any mutinous or seditious practices, or shall treat with contempt his superior, being in the execution of his office; or being witness to any mutiny or sedition, shall not do his utmost to suppress it, he shall be punished at the discretion of a court martial.

Art. XIV. No officer or private in the navy shall disobey the lawful orders of his superior officer, or strike him, or draw, or offer to draw, or raise any weapon against him, while in the execution of the duties of his office, on pain of death, or such other punishment as a court martial shall inflict.

Art. XV. No person in the navy shall quarrel with any other person in the navy, nor use provoking or reproachful words, gestures, or menaces, on pain of such punishment as a court martial shall adjudge.

Art. XVI. If any person in the navy shall desert to an enemy or rebel, he shall suffer death.

Art. XVII. If any person in the navy shall desert, or shall entice others to desert, he shall suffer death, or such other punishment as a court martial shall adjudge; and if any officer or other person belonging to the navy, shall receive or entertain any deserter from any other vessel of the navy, knowing him to be such, and shall not, with all convenient speed, give notice of such deserter to the commander of the vessel to which he belongs, or to the commander in chief, or to the commander of the squadron, he shall on conviction thereof, be cashiered, or be punished at the discretion of a court martial. All offences, committed by persons belonging to the navy while on shore, shall be punished in the same manner as if they had been committed at sea.

Art. XVIII. If any person in the navy shall knowingly make or sign, or shall aid, abet, direct, or procure the making or signing of any false muster, or shall execute, or attempt, or countenance any fraud against the United States, he shall, on conviction, be cashiered and rendered forever incapable of any future employment in the service of the United States, and shall forfeit all the pay and subsistence due him, and suffer such other punishment as a court martial shall inflict.

Art. XIX. If any officer, or other person in the navy, shall, through inattention, negligence, or any other fault, suffer any vessel of the navy to be stranded, or run upon rocks or shoals, or hazarded, he shall suffer such punishment as a court martial shall adjudge.

Art. XX. If any person in the navy shall sleep upon his watch, or negligently perform the duty assigned him, or leave his station before regularly relieved, he shall suffer death, or such punishment as a court martial shall adjudge; or, if the offender be a private, he may, at the discretion of the captain, be put in irons, or flogged not exceeding twelve lashes.

Art. XXI. The crime of murder, when committed by any officer, seaman, or marine, belonging to any public ship or vessel of the United States, without the territorial jurisdiction of the same, may be punished with death by the sentence of a court martial.

Art. XXII. The officers and privates of every ship or vessel, appointed as convoy to merchant or other vessels, shall diligently and faithfully discharge the duties of their appointment, nor shall they demand or exact any compensation for their services, nor maltreat any of the officers or crews of such merchant or other vessels, on pain of making such reparation as a court of admiralty may award, and of suffering such further punishment as a court martial shall adjudge.

Art. XXIII. If any commander or other officer shall receive or permit to be received, on board his vessel, any goods or merchandise, other than for the sole use of his vessel, except gold, silver, or jewels, and except the goods or merchandise of vessels which may be in distress, or shipwrecked, or in imminent danger of being

shipwrecked, in order to preserve them for their owner, without orders from the President of the United States or the navy department, he shall, on conviction thereof, be cashiered, and be incapacitated forever afterwards, for any place or office in the navy.

Art. XXIV. If any person in the navy shall waste, embezzle, or fraudulently buy, sell, or receive any ammunition, provisions, or other public stores; or if any officer or other person shall, knowingly, permit through design, negligence, or inattention, any such waste, embezzlement, sale or receipt, every such person shall forfeit all the pay and subsistence then due him, and suffer such further punishment as a court martial shall direct.

Art. XXV. If any person in the navy shall unlawfully set fire to or burn any kind of public property, not then in the possession of an enemy, pirate, or rebel, he shall suffer death: And if any person shall, in any other manner, destroy such property, or shall not use his best exertions to prevent the destruction thereof by others, he shall be punished at the discretion of a court martial.

Art. XXVI. Any theft not exceeding twenty dollars may be punished at the discretion of the captain, and above that sum, as a court martial shall direct.

Art. XXVII. If any person in the navy shall, when on shore, plunder, abuse, or maltreat any inhabitant, or injure his property in any way, he shall suffer such punishment as a court martial shall adjudge.

Art. XXVIII. Every person in the navy shall use his utmost exertions to detect, apprehend, and bring to punishment all offenders, and shall at all times, aid and assist all persons appointed for this purpose, on pain of such punishment as a court martial shall adjudge.

Art. XXIX. Each commanding officer shall, whenever a seaman enters on board, cause an accurate entry to be made in the ship's books, of his name, time, and term of his service; and before sailing transmit to the Secretary of the Navy, a complete list or muster roll of the officers and men under his command, with the date of their entering, time and terms of their service annexed; and shall cause similar lists to be made out on the first day of very second month, to be transmitted to the Secretary of the Navy, as opportunities shall occur; accounting in such lists or muster rolls, for any casualties which may have taken place since the last list or muster roll. He shall cause to be accurately minuted on the ship's books, the names of, and times at which any death or desertion may occur; and in case of death, shall take care that the purser secure all the property of the deceased for the benefit of this legal representative or representatives. He shall cause frequent inspections to be made into the condition of the provisions, and use every precaution for its preservation. He shall, whenever he orders officers and men to take charge of a prize, and proceed to the United States, and whenever officers or men are sent from his ship from whatever cause, take that each man be furnished with a complete statement of his account, specifying the date of his enlistment, and the period and terms of his service; which account shall be signed by the commanding officer and purser. He shall cause the rules for the government of the navy to be hung up in some public part of the ship, and read once a month to his ship's company. He shall cause a convenient place to be set apart for sick or disabled men, to which he shall have them removed, with their hammocks and bedding, when the surgeon shall so advise, and shall direct that some of the crew attend them and keep the place clean; and if necessary, shall direct that cradles, and buckets with covers, be made for their use: And when his crew is finally paid off, he shall attend in person, or appoint a proper officer, to see that justice be done to the men, and to the United States, in the settlement of the accounts. Any commanding officer, offending herein, shall be punished at the discretion of a court martial.

Art. XXX. No commanding officer shall, of his own authority, discharge a commissioned or warrant officer, nor strike nor punish him otherwise than by suspension or confinement, nor shall he of his own authority, inflict a punishment on any private beyond twelve lashes with a cat-of-nine-tails, nor shall he suffer any wired, or other than a plain, cat-of-nine-tails, to be used on board his ship; nor shall any officer who may command by accident, or in the absence of the commanding officer (except such commander be absent for a time by leave) order or inflict any other punishment than confinement, for which he shall account on the return of such absent commanding officer. Nor shall any commanding officer receive on board any petty officers or men turned over from any other vessel to him, unless each of such officers and men produce to him an account signed by the captain and purser of the vessel from which they came, specifying the date of such officer's or man's entry, the period and terms of service, the sums paid and the balance due him, and the quality in which he was rated on board such ship. Nor shall any commanding officer, having received any petty officer or man as aforesaid, rate him in a lower or worse station than that in which he formerly served. Any commanding officer offending herein, shall be punished at the discretion of a court martial.

Art. XXXI. Any master at arms, or other person of whom the duty of master at arms is required, who shall refuse to receive such prisoners as shall be committed to his charge, or having received them, shall suffer them to escape, or dismiss them without orders from proper authority, shall suffer in such prisoners' stead, or be punished otherwise at the discretion of a court martial.

Art. XXXII. All crimes committed by persons belonging to the navy, which are not specified in the foregoing articles, shall be punished according to the laws and customs in such cases at sea.

Art. XXXIII. All officers, not holding commissions or warrants, or who are not entitled to them, except such as are temporarily appointed to the duties of a commissioned or warrant officer, are deemed petty officers.

Art. XXXIV. Any person entitled to wages or prize money, may have the same paid to his assignee, provided the assignment be attested by the captain and purser; and in case of the assignment of wages, the power shall specify the precise time they commence. But the commander of every vessel is required to discourage his crew from selling any part of their wages or prize money, and never to attest any power of attorney, until he is satisfied that the same is not granted in consideration of money given for the purchase of wages or prize money.

Naval General Courts Martial.

Art. XXXV. General courts martial may be convened as often as the President of the United States, the Secretary of the Navy, or the commander in chief of the fleet, or commander of a squadron, while acting out of the United States, shall deem it necessary: *Provided,* that no general court martial shall consist of more than thirteen, nor less than five members, and as many officers shall be summoned on every such court as can be convened without injury to the service, so as not to exceed thirteen, and the senior officer shall always preside, the others ranking agreeably to the date of their commissions; and in no case, where it can be avoided without injury to the service, shall more than one half the members, exclusive of the president, be junior to the officer to be tried.

Art. XXXVI. Each member of the court, before proceeding to trial, shall take the following oath or affirmation, which the judge advocate or person officiating as such, is hereby authorized to administer.

"I, *A. B.* do swear [or affirm] that I will truly try, without prejudice or partiality, the case now depending, according to the evidence which shall come before the court, the rules for the government of the navy, and my own conscience; and that I will not by any means divulge or disclose the sentence of the court, until it shall have been approved by the proper authority, nor will I at any time divulge or disclose the vote or opinion of any particular member of the court, unless required so to do before a court of justice in due course of law."

This oath or affirmation being duly administered, the president is authorized and required to administer the following oath or affirmation to the judge advocate, or person officiating as such.

"I, *A. B.* do swear [or affirm] that I will keep a true record of the evidence given to and the proceedings of this court; nor will I divulge or by any means disclose the sentence of the court until it shall have been approved by the proper authority; nor will I at any time divulge or disclose the vote or opinion of any particular member of the court, unless required so to do before a court of justice in due course law."

Art. XXXVII. All testimony given to a general court martial shall be on oath or affirmation, which the president of the court is hereby authorized to administer, and if any person shall refuse to give his evidence as aforesaid, or shall prevaricate, or shall behave with contempt to the court, it shall and may be lawful for the court to imprison such offender at their discretion; provided that the imprisonment in no case shall exceed two months: and every person who shall commit wilful perjury on examination on oath or affirmation before such court, or who shall corruptly procure, or suborn any person to commit such wilful perjury, shall and may be prosecuted by indictment or information in any court of justice of the United States, and shall suffer such penalties as are authorized by the laws of the United States in case of perjury or the subornation thereof. And in every prosecution for perjury or the subornation thereof under this act, it shall be sufficient to set forth the offence charged on the defendant, without setting forth the authority by which the court was held, or the particular matters brought or intended to be brought before the said court.

Art. XXXVIII. All charges, on which an application for a general court martial is founded, shall be exhibited in writing to the proper officer, and the person demanding the court shall take care that the person accused be furnished with a true copy of the charges, with the specifications, at the time he is put under arrest, nor shall any other charge or charges, than those so exhibited, be urged against the person to be tried before the court, unless it appear to the court that intelligence of such charge had not reached the person demanding the court, when the person so to be tried was put under arrest, or that some witness material to the support of such charge, who was at that time absent, can be produced; in which case, reasonable time shall be given to the person to be tried to make his defence against such new charge. Every officer so arrested is to deliver up his sword to his commanding officer, and to confine himself to the limits assigned him, under pain of dismission from service.

Art. XXXIX. When the proceedings of any general court martial shall have commenced, they shall not be suspended or delayed on account of the absence of any of the members, provided five or more be assembled; but the court is enjoined to sit from day to day, Sun-

days excepted, until sentence be given: and no member of said court shall, after the proceedings are begun, absent himself therefrom, unless in case of sickness or orders to go on duty from a superior officer, on pain of being cashiered.

Art. XL. Whenever a court martial shall sentence any officer to be suspended, the court shall have power to suspend his pay and emoluments for the whole, or any part of the time of his suspension.

Art. XLI. All sentences of courts martial, which shall extend to the loss of life, shall require the concurrence of two-thirds of the members present; and no such sentence shall be carried into execution, until confirmed by the President of the United States; or if the trial take place out of the United States, until it be confirmed by the commander of the fleet or squadron: all other sentences may be determined by a majority of votes, and carried into execution on confirmation of the commander of the fleet, or officer ordering the court, except such as go to the dismission of a commissioned or warrant officer, which are first to be approved by the President of the United States.

A court martial shall not, for any one offence not capital, inflict a punishment beyond one hundred lashes.

Art. XLII. The President of the United States, or when the trial takes place out of the United States, the commander of the fleet or squadron, shall possess full power to pardon any offence committed against these articles, after conviction, or to mitigate the punishment decreed by a court martial.

Sec. 2. Art. I. *And be it further enacted,* That courts of inquiry may be ordered by the President of the United States, the Secretary of the Navy, or the commander of a fleet or squadron, provided such court shall not consist of more than three members who shall be commissioned officers, and a judge advocate, or person to do duty as such; and such courts shall have power to summon witnesses, administer oaths, and punish contempt in the same manner as courts martial. But such court shall merely state facts, and not give their opinion, unless expressly required so to do in the order for convening; and the party, whose conduct shall be the subject of inquiry, shall have permission to cross examine all the witnesses.

Art. II. The proceedings of courts of inquiry shall be authenticated by the signature of the president of the court and judge advocate, and shall, in all cases not capital, or extending to the dismission of a commissioned or warrant officer, be evidence before a court martial, provided oral testimony cannot be obtained.

Art. III. The judge advocate, or person officiating as such, shall administer to the members the following oath or affirmation:

"You do swear, [or affirm] well and truly to examine and inquire according to the evidence, into the matter now before you, without partiality or prejudice."

After which, the president shall administer to the judge advocate, or person officiating as such, the following oath or affirmation:

"You do swear [or affirm] truly to record the proceedings of this court, and the evidence to be given in the case in hearing."

Sec. 3. *And be it further enacted,* That in all cases, where the crews of the ships or vessels of the United States shall be separated from their vessels, by the latter being wrecked, lost or destroyed, all the command, power, and authority, given to the officers of such ships or vessels, shall remain and be in full force as effectually as if such ship or vessel were not so wrecked, lost, or destroyed, until such ship's company be regularly discharged from, or ordered again into the service, or until a court martial shall be held to inquire into the loss of such ship or vessel; and if by the sentence of such court, or other satisfactory evidence, it shall appear that all or any of the officers and men of such ship's company did their utmost to preserve her, and after the loss thereof behaved themselves agreeably to the discipline of the navy, then the pay and emoluments of such officers and men, or such of them as shall have done their duty as aforesaid, shall go on until their discharge or death; and every officer or private who shall, after the loss of such vessel, act contrary to the discipline of the navy, shall be punished at the discretion of a court martial, in the same manner as if such vessel had not been lost.

Sec. 4. *And be it further enacted,* That all the pay and emoluments of such officers and men, of any of the ships or vessels of the United States taken by an enemy, who shall appear by the sentence of a court martial, or otherwise, to have done their utmost to preserve and defend their ship or vessel, and, after the taking thereof, have behaved themselves obediently to their superiors, agreeably to the discipline of the navy, shall go on and be paid them until their death, exchange, or discharge.

Sec. 5. *And be it further enacted,* That the proceeds of all ships and vessels, and the goods taken on board of them, which shall be adjudged good prize, shall, when of equal or superior force to the vessel or vessels making the capture, be the sole property of the captors; and when of inferior force, shall be divided equally between the United States and the officers and men making the capture.

Sec. 6. *And be it [further] enacted,* That the prize money, belonging to the officers and men, shall be distributed in the following manner:

I. To the commanding officers of fleets, squadrons, or single ships, three twentieths, of which the commanding

officer of the fleet or squadron shall have one twentieth, if the prize be taken by a ship or vessel acting under his command, and the commander of single ships, two twentieths; but where the prize is taken by a ship acting independently of such superior officer, the three twentieths shall belong to her commander.

II. To sea lieutenants, captains of marines, and sailing masters, two twentieths; but where there is a captain, without a lieutenant of marines, these officers shall be entitled to two twentieths and one third of a twentieth, which third, in such case, shall be deducted from the share of the officers mentioned in article No. III. of this section.

III. To chaplains, lieutenants of marines, surgeons, pursers, boatswains, gunners, carpenters, and master's mates, two twentieths.

IV. To midshipmen, surgeon's mates, captain's clerks, schoolmasters, boatswain's mates, gunner's mates, carpenter's mates, ship's steward, sail-makers, masters at arms, armorers, cockswains, and coopers, three twentieths and an half.

V. To gunner's yeoman, boatswain's yeoman, quartermasters, quarter-gunners, sail-maker's mates, sergeants and corporals of marines, drummers, fifers, and extra petty officers, two twentieths and an half.

VI. To seamen, ordinary seamen, marines, and all other persons doing duty on board, seven twentieths.

VII. Whenever one or more public ships or vessels are in sight at the time any one or more ships are taking a prize or prizes, they shall all share equally in the prize or prizes, according to the number of men and guns on board each ship in sight.

No commander of a fleet or squadron shall be entitled to receive any share of prizes taken by vessels not under his immediate command; nor of such prizes as may have been taken by ships or vessels intended to be placed under his command, before they have acted under his immediate orders; nor shall a commander of a fleet or squadron, leaving the station where he had the command, have any share in the prizes taken by ships left on such station, after he has gone out of the limits of his said command.

SEC. 7. *And be it further enacted,* That a bounty shall be paid by the United States, of twenty dollars for each person on board any ship of an enemy at the commencement of an engagement, which shall be sunk or destroyed by any ship or vessel belonging to the United States of equal or inferior force, the same to be divided among the officers and crew in the same manner as prize money.

SEC. 8. *And be it further enacted,* That every officer, seaman, or marine, disabled in the line of his duty, shall be entitled to receive for life, or during his disability, a pension from the United States according to the nature and degree of his disability, not exceeding one half his monthly pay.

SEC. 9. *And be it [further] enacted,* That all money accruing, or which has already accrued to the United States from the sale of prizes, shall be and remain forever a fund for the payment of pensions and half pay, should the same be hereafter granted, to the officers and seamen who may be entitled to receive the same; and if the said fund shall be insufficient for the purpose, the public faith is hereby pledged to make up the deficiency; but if it should be more than sufficient, the surplus shall be applied to the making of further provision for the comfort of the disabled officers, seamen, and marines, and for such as, though not disabled, may merit by their bravery, or long and faithful services, the gratitude of their country.

SEC. 10. *And be it further enacted,* That the said fund shall be under the management and direction of the Secretary of the Navy, the Secretary of the Treasury, and the Secretary of War, for the time being, who are hereby authorized to receive any sums to which the United States may be entitled from the sale of prizes, and employ and invest the same, and the interest arising therefrom, in any manner which a majority of them may deem most advantageous. And it shall be the duty of the said commissioners to lay before Congress, annually, in the first week of their session, a minute statement of their proceedings relative to the management of said fund.

SEC. 11. *And be it further enacted,* That the act passed the second day of March, in the year one thousand seven hundred and ninety-nine, entitled "An act for the government of the navy of the United States," from and after the first day of June next, shall be, and hereby is repealed.

APPROVED, April 23, 1800.

Statutes at Large of the United States of America, 1789–1845, 8 vols. (Boston: Little, Brown and Company, 1846–67), 2:45–53. This act was Chapter XXXIII of the statutes passed during the first session of the sixth Congress.

U.S. Navy Regulations, 1814

NAVAL REGULATIONS,
ISSUED BY COMMAND
OF THE
PRESIDENT
OF THE
UNITED STATES OF AMERICA.

NAVAL REGULATIONS.

OF THE DUTIES OF COMMANDER OF A SQUADRON.

1. He is to inform the secretary of the navy of all his proceedings which relate to the service upon which he may be ordered.

2. He is to correspond with the public offices, about such matters as relate to them, and send to them an account of all directions given by him to those under his command, which concern the said offices.

3. In order that he may use the vessels of his squadron to the greatest advantage, as occasion may require, he is to inform himself of their qualities.

4. In order to facilitate the operations for which the squadron is destined, its commandant shall take care to distribute his orders to all the commanders under him, regulated by his instructions from the secretary of the navy.

5. Immediately on his receiving orders to sail, he shall weigh anchor as soon as the weather will permit; and previous to his departure, he shall give an account to the secretary of the navy of the condition of his squadron, without omitting any essential circumstance.

6. He shall suit his sails according to the qualities of the ships and circumstances of the weather, without obliging the heaviest sailers to an extraordinary exertion, from whence damage might result.

7. When the fleet shall be divided into squadrons or divisions, all the ships shall regulate their motions by those of their respective commandants.

8. The commandant shall always maintain his squadron in readiness to sail expeditiously: he shall from time to time visit the ships, as well to examine if they are in readiness, as to take care that they observe good discipline.

9. He may suspend from their stations the captains of vessels, or any other officers under his command, who, for bad conduct, or incapacity, he shall think deserving of such punishment; but he must immediately transmit an account thereof to the secretary of the navy, specifying his reasons for so doing, and furnish the captain or officer suspended with a copy thereof.

10. The commandant of the squadron ought not to alter the appointments assigned to the officers at the time of fitting out, without reasons.

11. He is to preserve the instructions and orders which he may receive from the navy office, and all other papers and correspondence relating to the service upon which he may be ordered, in the most intelligible form.

12. At the end of the cruise, he shall transmit to the secretary of the navy, a fair copy of all his official correspondence He is to deliver to the secretary of the navy his journal, which he is to make during the cruise with the greatest exactness.

13. He is never to give orders to any captain to bear supernumeraries, unless there be good cause for it, which is to be expressed in the body of the order; and he is to inform the secretary of the navy when he gives such orders, and of his reasons for so doing.

14. When he is at sea, he is frequently to exercise the ships under his command, and draw them into line of battle, when the weather is fair, and the same can be done consistently with his cruising orders, and without interruption to the voyage.

15. He is to visit the ships of his squadron or division, and view the men on board, and see them mustered as often as he shall think necessary.

16. When he is in foreign parts, where naval or other agents are established, he is to conform himself, as much as possible, to the standing rules of the navy, in such directions as he shall have occasion to give to them; and never to put them under any extraordinary expenses, unless the service should absolutely require the same.

17. He is never to interest himself in the purchase of any stores or provisions in foreign parts, where there are proper officers appointed for that service; except there shall be an absolute necessity to make use of his credit or authority, to procure such provisions or stores as are wanted; but in that case, he shall not be so concerned as to have any private interest in the same.

Of the duties of a Captain or Commander.

1. When a captain or commander is appointed to command one of the United States' ships, he is immediately to repair on board, and visit her throughout.

2. To give his constant attendance on board, and quicken the dispatch of the work; and to send to the navy department weekly accounts, or oftener, if necessary, of the condition and circumstances she is in, and the progress made in fitting her out.

3. To take inventories of all the stores committed to the charge of his officers respectively, and to require from his boatswain, gunner, sailmaker, carpenter and purser, counterparts of their respective indents.

4. To cause his clerk to be present, and to take an account of all the stores and provisions that come on board, and when; which account he is to compare with the indents, in order to prevent any fraud or neglect.

5. To keep counter-books of the expense of the ship's stores and provisions, whereby to know the state and condition of the same; and to audit the accounts of the officers entrusted therewith, once a week, in order to be a check upon them.

6. When ordered to recruit, he is to use his best endeavours to get the ship manned, and not to enter any but men of able bodies, and fit for service: he is to keep the established number of men complete, and not to exceed his complement.

7. When the ship's company is completed, they shall be divided into messes and guards; and he shall order, without delay, the partition of the people for an engagement, to the end that, before they sail, every one may know his post.

8. He may grant to private ships of the nation the succours he lawfully may, taking from their captains or patrons a correspondent security, that the owners may satisfy the amount or value of the things supplied.

9. At all times, whether sailing alone or in a squadron, he shall have his ship ready for an immediate engagement: to which purpose, he shall not permit any thing to be on deck that may embarrass the management of the guns, and not be readily cleared away.

10. As, from the beginning of the cruise the plan of the combat ought to be formed, he shall have his directions given, and his people so placed, as not to be unprovided against any accident which may happen.

11. If it is determined to board the enemy, the captain is not, under any pretext, to quit his ship, whose preservation must be the chief object of his care; but he may appoint his second in command, or any other officer he thinks proper for that duty, without attending to rank.

12. He shall observe, during his cruise, the capacity, application, and behavior of his officers; and to improve them, he shall employ them in works and commissions that may manifest their intelligence.

13. He is to cause all new-raised men and others, not skilled in seamanship, daily to lash up their hammocks, and carry them to the proper places for barricading the ship, whenever the weather will permit; and also to have them practised in going frequently every day up and down the shrouds, and employed on all kinds of work, to be created purposely to keep them in action, and to teach them the duty of seamen.

14. To keep a regular muster-book, setting down therein the names of all persons entered to serve on board, with all circumstances relating to them.

15. Himself to muster the ship's company at least once a week, in port or at sea, and to be very exact in this duty; and if any person shall absent himself from his duty, without leave, for three successive musters, he is to be marked as a run-away, on the ship's books.

16. To send, every month, one muster-book complete, to the navy office, signed by himself and purser.

17. To make a list of seamen run away, inserting the same at the end of the muster-books, and to distinguish the time, manner and by what opportunity they made their escape: if the desertion happens in any port of the United States, he is to send to the navy department their names, place of abode, and all the circumstances of their escape.

18. The captain of the ship shall be responsible for his crew, whose desertion shall be laid to his charge, whenever it proceeds from a want of necessary care; but if it proceeds from the neglect of an officer who shall have the charge of a watering party, or any other duty on shore, and, from his negligence, any part of the crew entrusted to him shall desert, that officer shall be responsible for the same.

19. He is to make out tickets for all such seamen as shall be discharged from his books, signed by himself and purser, and to deliver them to none but the party; and if the party be dead or absent, he is to send the ticket forthwith to the navy-office.

20. He is not to suffer the ship's stores to be misapplied or wasted; and if such loss happens by the negligence or wilfulness of any of the ship's company, he is to charge the value thereof against the wages of the offender, on the muster and pay-books.

21. He shall make no alteration in any part of the ship.

22. He is to keep sentinels posted at the scuttle, leading into all the store-rooms, and no person is to pass down but by leave from the captain or commanding officer of the watch, which leave must be signified to the sentinel from the quarter-deck.

23. He is to observe seasonable times in setting up his shrouds and other rigging, especially when they are new and apt to stretch; and also to favor his masts as much as possible.

24. He is to cause such stores as require it, to be frequently surveyed and aired, and their defects repaired; and the store-rooms to be kept airy and in good condition, and secured against rats.

25. He is not to make use of ship's sails for covering boats, or for awnings.

26. The decks or gratings are not to be scraped oftener than is necessary, but are to be washed and swabbed once a day, and air let into the hold as often as may be.

27. He is to permit every officer to possess his proper cabin, and not to make any variation therein.

28. No person is to lie upon the orlop but by leave from the captain, nor to go amongst the cables with candles, but when service requires it.

29. Such as smoke tobacco are to take it in the forecastle, and in no other place, without the captain's permission, which is never to be given to smoke below the upper gun-deck.

30. Care is to be taken every night, on setting the watch, that all fire and candles be extinguished in the cook-room, hold, steward-room, cock-pit, and every where between decks; nor are candles to be used in any other part of the ship but in lanthorns, and that not without the captain's leave; and the lanthorns must always be whole and unbroken.

31. He is not to suffer any person to suttle or sell any sorts of liquors to the ship's company, nor any debts for the same to be inserted in the slop-book, on any pretence whatsoever.

32. Before the ship proceeds to sea, he is, without any partiality or favor, to examine and rate the ship's company, according to their abilities, and to take care that every person in the ship, without distinction, do actually perform the duty for which he is rated.

33. Before the ship sails, he is to make a regulation for quartering the officers and men, and distributing them to the great guns, small-arms, rigging, &c.; and a list of such order and distribution is to be fixed up in the most public place of the ship. He is also frequently to exercise the ship's company in the use of the great guns and small-arms; and to set down in his journal the times he exercises them.

34. The following number of men at least, (exclusive of marines) are to be exercised and trained up to the use of small-arms, under the particular care of a lieutenant or master at arms.

44 gun ship, - - - - - - - - - - -	75 men.
36 do. - - - - - - - - - - - - -	60 do.
32 do. - - - - - - - - - - - - -	45 do.
24 and under 32 gun ships, - - - -	40 do.
18 and under 24 do. - - - - - - -	30 do.
All smaller vessels, - - - - - - - -	20 do.

35. If any officers are absent from their duty when the ship is under sailing orders, he is to send their names to the navy-office, with the cause of their absence.

36. He is to take care of his boats and secure them before blowing weather; also, the colors are not to be kept abroad in windy weather, but due care taken of them.

37. He is not to carry any woman to sea, without orders from the navy office, *or the commander of the squadron.*

38. When he is to sail from port to port in time of war, or appearance thereof, he is to give notice to merchantmen bound his way, and take them under his care, if they are ready; but not to make unnecessary stay, or deviate from his orders on that account.

39. He is to keep a regular journal, and at the expiration of the voyage, to give in a general copy to the navy-office.

40. He is, by all opportunities, to send an account of his proceedings to the navy-office, with the condition of the ship, men, &c.; he is likewise to keep a punctual correspondence with every of the public offices, in whatsoever respectively concerns them.

41. He is not to go into any port, but such as are directed by his orders, unless necessitously obliged, and then not to make any unnecessary stay; if employed in cruising, he is to keep the sea the time required by his orders, or give reasons for acting to the contrary.

42. Upon all occasions of anchoring, he is to take great care in the choice of a good birth, and examine the quality of the ground for anchoring, where he is a stranger, sounding at least three cables lengths round the ship.

43. In foreign ports he is to use the utmost good husbandry in careening the ship, and not to do it but under an absolute necessity; none are to be employed in careening and refitting the ship but the ship's company, where it can be avoided; and for the encouragement of his own men, they are entitled to an extraordinary allowance per day; and to prevent any abuse herein, each ship has the number of operative men limited as follows:

	In the United States.	In all foreign parts.
To master carpenters, carpenters' mates, shipwrights and caulkers, for working on board the ship they belong to, in caulking and fitting her for careen, and graving or tallowing her, per day,	50 cents.	75 cents.
For working on board any other of the United States' ships.	75 cents.	1 dollar.

And there shall be allowed no more for caulking a ship, fitting her for careen, graving or tallowing her, or other necessary works for each careening or cleaning, than what amounts to the labor of the following number of men for one day, viz.

For a 44 – 180 men for one day.		
For a 36 – 160 do.		do.
For a 32 – 140 do.		do.
For a 24 – 90 do.		do.
For an 18 – 70 do.		do.
All under – 30 do.		do.

44. If he is obliged to take up money abroad for the use of the ship, he is to negotiate it at the best exchange.

45. He is to advise the proper officer of what bills he draws, with the reasons thereof, and with the said bills send duplicates of his accounts, and vouchers for his disbursements, signed by himself and purser.

46. He is to take care that all stores brought on board, be delivered to the proper officers; and to take their receipts for the same.

47. Upon the death of any officer, he is to take care that an inventory be taken of all his goods and papers, and that the same be sealed up, and reserved for the use of such as have a legal right to demand them.

48. When any officer who has the custody of stores or provisions shall die, be removed or suspended, he is to cause an exact survey and inventory to be taken forthwith of the remains of such stores, which is to be signed by the successor, who is to keep a duplicate thereof, and also by the surveying officers.

49. Upon his own removal into another ship, he is to show the originals of all such orders as have been sent to him, and remain unexecuted, to his successor, and leave with him attested copies of the same.

50. He is to leave with his successor a complete muster-book, and send up all other books and accounts under his charge, to the officers they respectively relate to.

51. In case of shipwreck, or other disaster, whereby the ship may perish, the officers and men are to stay with the wreck as long as possible, and save all they can.

52. When any men borne for wages are discharged from one ship to another, the captain of the ship from which they are so discharged, is to send immediately pay-lists for such men to the navy-office, and the purser of the ship from which they are so discharged, is also to supply the purser of the ship to which they are transferred, a pay-list, stating the balances respectively due them.

53. To promote cleanliness and health, the following rules are to be attended to. 1. All men on board are to keep themselves in every respect as clean as possible. 2. That the ship be aired between decks as much as may be, and that she be always kept thoroughly clean. 3. That all necessary precautions be used, by placing sentinels or otherwise, to prevent people easing themselves in the hold, or throwing any thing there that may occasion nastiness. 4. That no fruit or strong liquors be sold on board the ship; *except in the judgment of the commander of the squadron, a limited quantity of fruit be necessary for the health of the crew, in which case he will issue an order.*

54. He is responsible for the whole conduct and good government of the ship, and for the due execution of all regulations which concern the several duties of the officers and company of the ship, who are to obey him in all things which he shall direct them for the service of the United States.

55. He is answerable for the faults of his clerk; nor can he receive his wages without the proper certificates, and must make good all damages sustained by his neglect or irregularity.

56. The quarter-deck must never be left without one commissioned officer, at least, *and the other necessary officers which the captain may deem proper,* to attend to the duty of the ship.

57. Commanding officers are to discourage seamen from selling their wages; and not to attest letters of attorney, if the same appear granted in consideration of money given for the purchase of wages.

Of the duties of a Lieutenant.

1. He shall promptly, faithfully, and diligently, execute all such orders as he shall receive from his commander, for the public service, nor absent himself from the ship without leave, on any pretence.

2. He is to keep a list of the officers and men in his watch, muster them, and report the names of the absentees. He is to see that good order be kept in his watch, that no fire or candle be burning, and that no tobacco be smoked between decks.

3. He is not to change the course of the ship at sea without the captain's directions, unless to prevent an immediate danger.

4. No boats are to come on board or go off without the lieutenant of the watch being acquainted with it.

5. He is to inform the captain of all irregularities, and to be upon deck in his watch, and prevent noise or confusion.

6. He is to see that the men be in their proper quarters in time of action; and that they perform all their duty.

7. The youngest lieutenant is frequently to exercise the seamen in the use of small-arms; and in the time of action he is to be chiefly with them.

8. He is to take great care of the small-arms, and see that they be kept clean, and in good condition for service, and that they be not lost or embezzled.

9. The first lieutenant is to make out a general alphabetical book of the ship's company, and proper watch, quarter and station bills, in case of fire, manning of ship, loosing and furling of sails, reefing of topsails at sea, working of ship, mooring and unmooring, &c. leaving room for unavoidable alterations. This is to be hung in some public part of the ship, for the inspection of every person concerned.

10. No lieutenant, or other officer, belonging to a ship of the United States, to go on shore, or on board another vessel, without first obtaining permission from the captain or commanding officer, on his peril; and in the absence of the captain, the commanding officer to grant no permission of this sort, without authority from the captain, previous to the captain's leaving the ship.

Of the duties of a Sailing Master.

1. He is to inspect the provisions and stores sent on board, and of what appears not good, he is to acquaint the captain.

2. He is to take care of the ballast, and see that it be clean and wholesome, and sign for the quantity delivered; and, in returning ballast, to see that vessels carry away their full lading.

3. He is to give his directions in stowing the hold, for the mast-room, trimming the ship, and for preservation of the provisions; and the oldest provisions to be stowed, so as to be first expended.

4. He is to take special care that the rigging and stores be duly preserved; and to sign the carpenter's and boatswain's expense-book, taking care not to sign undue allowances.

5. He is to navigate the ship under the direction of his superior officer, and see that the log and log-book be duly kept, and to keep a good look-out.

6. He is duly to observe the appearances of coasts; and if he discovers any new shoals, or rocks under water, to note them down in his journal, with their bearing and depth of water.

7. He is to keep the hawser clear when the ship is at anchor, and see that she is not girt with her cables.

8. He is to provide himself with proper instruments, and books of navigation.

9. He is to be very careful not to sign any accounts, books, lists, or tickets, before he has thoroughly informed himself of the truth of every particular contained in the same.

10. He is to keep the ship in constant trim, and frequently to note her draught of water in the log-book. He is to observe the alterations made by taking in stores, water or ballast; and when the ship is in chase, or trying her sailing with another, he is to make memorandums of the draughts of water, the rake of the masts, state of the rigging, and to note every possible observation, that may lead to the knowledge of the ship's best point of sailing.

Of the duties of a Surgeon.

1. To inspect and take care of the necessaries sent on board for the use of the sick men; if not good, he must acquaint the captain; and he must see that they are duly served out for the relief of the sick.

2. To visit the men under his care twice a day, or oftener, if circumstances require it: he must see that his mates do their duty, so that none want due attendance and relief.

3. In cases that are difficult, he is to advise with the surgeons of the squadron.

4. To inform the captain daily of the state of his patients.

5. When the sick are ordered to the hospitals, he is to send with them to the surgeon, an account of the time and manner of their being taken ill, and how they have been treated.

6. But none are to be sent to sick-quarters, unless their distempers, or the number of the sick on board, are such that they cannot be taken due care of; and this the surgeon is to certify under his hand, before removal.

7. To be ready with his mates and assistants in an engagement, having all things at hand necessary for stopping of blood and dressing of wounds.

8. To keep a day-book of his practice, containing the names of his patients, their hurts, distempers, when taken ill, when recovered, removal, death, prescriptions, and method of treatment, while under cure.

9. From the last book he is to form two journals, one containing his physical, and the other his chirurgical practice.

10. Stores for the medical department are to be furnished upon his requisition; and he will be held responsible for the expenditure thereof.

11. He will keep a regular account of his receipts and expenditures of such stores, and transmit an account thereof to the accountant of the navy, at the end of every cruise.

Of the duties of a Chaplain.

1. He is to read prayers at stated periods; perform all funeral ceremonies over such persons as may die in the service, in the vessel to which he belongs: or, if directed by the commanding officer, over any person that may die in any other public vessel.

2. He shall perform the duty of a schoolmaster; and to that end, he shall instruct the midshipmen and volunteers, in writing, arithmetic and navigation, and in

whatsoever may contribute to render them proficients. He is likewise to teach the other youths of the ship, according to such orders as he shall receive from the captain. He is to be diligent in his office.

Of the duties of a Boatswain and Master Sail Maker.

1. THE boatswain is to receive into his charge the rigging, cables, cordage, anchors, sails, boats, &c.

2. He is not to cut up any cordage or canvass without an order in writing from the captain, and under the inspection of the master; and always to have by him a good quantity of small plats for security of the cables.

3. He and his mates are to assist and relieve the watch, see that the men attend upon deck, and that the working of the ship be performed with as little confusion as may be.

4. His accounts are to be audited and vouched by the captain and master, and transmitted to the navy-office.

5. If he has cause of complaint against any of the officers of the ship, with relation to the disposition of the stores under his charge, he is to represent the same to the navy-office, before the pay of the ship. He is not to receive his own wages until his accounts are passed.

6. He is not to sign any accounts, books, lists, or tickets, before he has thoroughly informed himself of the truth of every particular therein contained.

7. *Master Sail-maker.* He is, with his mate and crew, to examine all sails that are brought on board, and to attend all surveys and conversions of sails.

8. He is always, and in due time, to repair and keep the sails in order, fit for service.

9. He is to see that they are dry when put into the store-room, or very soon to have them taken up and aired, and see that they are secured from drips, damps and vermin, as much as possible.

10. When any sails are to be returned into store, he is to attend the delivery of them for their greater safety.

Of the duties of a Gunner, Armorer, and Gunsmith.

1. THE gunner is to receive, by indenture, the ordnance, ammunition, small-arms, and other stores allowed for the voyage; and if any part thereof be not good, he is to represent the same to the captain, in order to its being surveyed and returned.

2. He is to see that the powder-room be well secured, and in right order, before the powder is brought into the ship.

3. Powder in the copper-hooped barrels to be lodged in the ground tier; to see that the doors of the powder-room be fast locked, the skuttle well shut and covered, and to deliver the keys to the captain.

4. He is timely to advise the captain when any powder comes on board, nor is he to remove it, prepare furzes [*fuses*], &c. without the captain's directions, so that the fire and candles may be extinguished, sentinels posted, and all care used to prevent accidents.

5. He is not to go or send any one into the powder-rooms, but by leave of the captain, and to take care that they have nothing about them that will strike fire in falling.

6. No more than three rounds of parchment cartridges are to be filled at a time.

7. Perishing stores are to be surveyed and condemned; but if near any port in the United States, and they can conveniently be returned into store, they must be, otherwise may be thrown overboard.

8. Empty powder-barrels are not be staved, but preserved, to shift such as may be decayed.

9. *The Armorer and Gunsmith* are to assist the gunner in the survey and receipt of small-arms, and to keep them clean and in good order; but not to take them too often to pieces, which is detrimental to locks, &c.

10. Their station is in the gun-room, or such other place as the commanding officer may direct, where they are to observe the gunner's orders.

11. *The Gunner* is to receive the armorer's tools, and to account for them at the end of the voyage, in the same manner as for the other stores under his charge.

12. In foreign parts, if the small-arms want such repairs as cannot be done on board, the captain must cause a survey, and the defectives may be sent ashore to be repaired; but the armorer or gunsmith must attend to see the reparations well executed. They must return the small-arms into store, clean and in good order.

13. The quantities of powder for exercise, and on occasions of service and scaling, must be regulated by the captain or commanding officer. In time of action the allowance of powder must be reduced by degrees, until the same be lessened to one fourth of the weight of the shot. He is not to swab a gun when it grows hot, for fear of splitting.

14. He is to take care that the guns be placed upon their proper carriages; for by this means they will fit, and stand a proper height for the sill of the ports.

15. He is not to scale the guns oftener than the ship is refitted, unless upon extraordinary occasions, and with the captain's orders; and when they are loaded for service, he is to see them well tompioned, and the vents filled with oakum.

16. He is to use great caution in order to prevent damage to such guns as are struck into the hold, by paying them all over, with a coat of warm tar and tallow mixt, &c.

17. He is to take care of the stores committed to him; for no waste that is not perishable, will be allowed him, only reasonable wear; and if any accident, it must be vouched by the captain.

18. He is to keep the boxes of grape-shot and hand-grenadoes in a dry place.

19. He is not to load the guns with unfixt mixtures, which greatly endanger their splitting.

20. If he has cause of complaint against any of the officers of the ship, with relation to the disposition of the stores under his charge, he is to represent the same to the navy-office, before the pay of the ship.

Of the duties of a Carpenter.

1. To take upon him the care and preservation of the ship's hull, masts, &c.; and also the stores committed to him by indenture.

2. To visit and inspect all parts of the ship daily; to see that all things are well secured and caulked; order the pumps and make report to the captain.

3. In an engagement, he is to be watchful, and have all materials ready to repair damages; and frequently to pass up and down the hold with his crew, to be ready to plug up shot-holes.

Of the duties of a Master-at-arms and Corporal.

1. Daily, by turns, (as the captain shall appoint) to exercise the ship's company.

2. He is to place and relieve sentinels, to mount with the guard, and to see that the arms be kept in order.

3. He is to see that the fire and candles be put out in season, and according to the captain's order.

4. He is to visit all vessels coming to the ship, and prevent the seamen going from the ship, without leave.

5. He is to acquaint the officer of the watch with all irregularities in the ship, which shall come to his knowledge.

6. *The Corporals*, are to act in subordination to the master-at-arms, and to perform the same duty under him, and to perform the duty themselves, where a master-at-arms is not allowed.

Of the duties of Midshipmen.

1. No particular duties can be assigned to this class of officers.

2. They are promptly and faithfully to execute all the orders for the public service, of their commanding officers.

3. The commanding officers will consider the midshipmen as a class of officers meriting, in an especial degree, their fostering care. They will see, therefore, that the schoolmasters perform their duty towards them, by diligently and faithfully instructing them in those sciences appertaining to their department; that they use their utmost care to render them proficients therein.

4. Midshipmen are to keep regular journals, and deliver them to the commanding officer at the stated periods, in due form.

5. They are to consider it as the duty they owe to their country, to employ a due portion of their time in the study of naval tactics, and in acquiring a thorough and extensive knowledge of all the various duties to be performed on board of a ship of war.

Of the duties of a Cook.

1. He is to have charge of the steep-tub, and is answerable for the meat put therein.

2. He is to see the meat duly watered, and the provisions carefully and cleanly boiled, and delivered to the men, according to the practice of the navy.

3. In stormy weather he is to secure the steep-tub, that it may not be washed overboard; but if it should be inevitably lost, the captain must certify it, and he is to make oath to the number of pieces so lost, that it may be allowed in the purser's account.

There shall be a distinct apartment appropriated on board of each vessel, for the surgeon, purser, boatswain, gunner, sailmaker, and carpenter, that they may keep the public goods committed respectively to their care.

Regulations to be observed respecting Provisions.

1. Provisions and slops are to be furnished upon the requisitions of the commanding officer, founded upon the purser's indents.

2. The purser being held responsible for the expenditure, shall, as far as may be practicable, examine and inspect all provisions offered to the vessel; and none shall be received that are objected to by him, unless they are examined and approved of, by at least two commissioned officers of the vessel.

3. In all cases where it may appear to the purser that provisions are damaged or spoiling, it will be his duty to apply to the commanding officer, who will direct a survey, by three officers, one of whom, at least, to be commissioned.

4. If upon a settlement of the purser's provision account, there shall appear a loss or deficiency of more than seven and a half per cent upon the amount of provisions received, he will be charged with, and held responsible for, such loss or deficiency, exceeding the seven and half per cent, unless he shows, by regular surveys, that the loss has been unavoidably sustained by damage or otherwise.

5. Captains may shorten the daily allowance of provisions when necessity shall require it, taking due care that each man has credit for his deficiency, that he may be paid for the same.

6. No officer is to have whole allowance while the company is at short.

7. Beef for the use of the navy is to be cut into ten pound pieces, pork into eight pound; and every cask to have the contents thereof marked on the head, and the person's name by whom the same was furnished.

8. If there be a want of pork, the captain may order beef in the proportion established, to be given out in lieu thereof, and *vice versa.*

9. One half gallon of water at least shall be allowed every man in foreign voyages, and such further quantity as shall be thought necessary on the home station; but on particular occasions the captain may shorten this allowance.

10. To prevent the buying of casks abroad, no casks are to be shipped which will want to be replaced by new ones, before the vessel's return to the United States.

11. If any provisions slip out of the slings, or are damaged through carelessness, the captain is to charge the value against the wages of the offender.

12. Every ship to be provided with a seine, and the crew supplied with fresh provisions as it can conveniently be done.

Regulations respecting Slops.

1. Slop-clothing is to be charged to the purser at the cost and charges; and he is to be held accountable for the expenditure.

2. And in no case will the purser be credited even for any alleged loss by damage in slops, unless he shows by regular surveys signed by three officers, one of whom at least to be commissioned, that the loss has been unavoidably sustained by *damage,* and not by any neglect or inattention on his part.

3. And, as a compensation for the risque and responsibility, the purser shall be authorised to dispose of the slops to the crew at a profit of five per cent; but he must, at the end of every cruise render a regular and particular slop-account, showing by appropriate columns the quantities of each several kind of articles received or purchased, and the prices and amount, and from whom, when and where; and he shall show the quantities disposed of, and to whom, and at what prices; so that his slop-account will show the articles, prices, and amount, received and disposed of.

4. On the death or removal of a purser, the commanding officer will cause a regular survey to be made on the slops remaining on hand, and an inventory thereof to be made out and signed by at least two commissioned officers.

5. Seamen, destitute of necessaries, may be supplied with slops by an order from the captain, after the vessel has commenced her voyage.

6. None are to receive a second supply until they have served full two months, and then not exceeding half their pay, and in the same proportion for every two months, if they shall be in want.

7. Slops are to be issued out publicly and in the presence of an officer, who is to be appointed by the captain, to see the articles delivered to the seamen and others, and the receipts given for the same, which he is also to certify.

8. The captain is to oblige those who are ragged, or want bedding, to receive such necessaries as they stand in need of.

9. The captain is to sign the slop-book before the ship is paid off; or on his removal from the ship at any time, the purser is to send the same to the proper accounting officer, duly signed.

10. On the discharge of a man by ticket, the value of the clothes he has been supplied with, must be noted on the same in words at length.

11. If necessity requires the buying of clothes in foreign parts, the captain must cause them to be procured of the kinds prescribed for the navy, and as moderate as possible: he must also, by the first opportunity, cause an invoice of the same to be forwarded to the navy department.

Regulations respecting the form and mode of keeping the Log-book and Journals on board of ships, or other vessels, of the United States.

For the purpose of establishing uniformity, the President orders as follows, viz.

1. The quarter-bill, log-tables or book, and journals of the officers, must be kept conformably to the annexed models. (*See following page.*)

2. The captains or commanders will cause to be laid before them, the first and fifteenth of every month, the journals of the sea lieutenants, masters, midshipmen, and volunteers under their orders, and will examine and compare them with their own.

3. If any of the said journals contain observations or remarks which may contribute to the improvement of geography, by ascertaining the latitude and longitude, fixing or rectifying the position of places, the heights and views of land, charts, plans or descriptions of any port, anchorage ground, coasts, islands, or danger little known; remarks relative to the direction and effects of currents, tides or winds: the officers or persons appointed to examined them, will make extracts of whatever

MODEL OF A JOURNAL, kept on board the United States of guns Commander, by

H	K	F	Courfes	Winds.	Occurrences, remarks and hiftorical events, &c made on board the United States of guns Commander, on the day of year	Refult of Day's work.
1						
2						Courfe made good.
3						
4						Diftance.
5						
6						Diff. latt'de.
7						
8						Departure.
9						
10						Mer'd. diftance.
11						
12						DD long'de.
1						
2						Long'de. ob'd.
3						
4						Latt'de ob'd.
5						
6						Var'n pr. amp'de
7						
8						Var'n pr. azim'th.
9						
10						Current.
11						
12					Diftance per Log—	

MODEL OF A LOG-BOOK, kept on board the United States of guns, Commander, by Sailing Mafter.

H	K	F	Courfes	Winds.	Occurrences and remarks, on board the United States Frigate of guns Commander, on the day of year
1					
2					
3					
4					
5					
6					
7					
8					
9					
10					
11					
12					
1					
2					
3					
4					
5					
6					
7					
8					Latt'de. Obf'd
9					Long'de. Obf'd
10					Vari. Even'g. Amp'de.
11					Diftance per Log— Vari. Morn'g. Amp'de.
12					

appears to merit to be preserved; and after these extracts have been communicated to the officer or author of the journal from which they have been drawn, and that he has certified in writing to the fidelity of his journal, as well as of the charts, plans and views, which he has joined to it, the same shall be signed by the officers and examiners, and transmitted with their opinion thereon to the secretary of the navy, to be preserved in the depot of charts, plans, and journals.

Regulations respecting Courts Martial.

1. ALL courts martial are to be held, offences tried, sentences pronounced, and execution of such sentences done, agreeably to the articles and orders contained in an act of Congress, made on the 23d of April, in the year 1800, entitled, "An act for the better government of the navy of the United States."

2. Courts martial may be convened as often as the President of the United States, the secretary of the navy, or commander in chief of a fleet, or commander of a squadron, while acting out of the United States, shall deem it necessary.

3. All complaints are to be made in writing, in which are to be set forth the facts, time, place, and the manner how they were committed.

4. The judge advocate is to examine witnesses upon oath, and by order of the commander in chief, or, in his absence, of the president of the court, to send an attested copy of the charge to the party accused, in time to admit his preparing his defence.

5. In all cases, the youngest member must vote first, and so proceed up to the president.

Regulations respecting Convoys.

1. A COMMANDER of a squadron, or commander of a ship appointed to convoy the trade of the United States, must give necessary and proper instructions in writing, and signed by himself, to all the masters of merchant ships and vessels under his protection.

2. He is to take an exact list, in proper form, containing the names of all the ships and vessels under his convoy, and send a copy thereof to the navy department, before he sails.

3. He is not, in time of actual war, to chase out of sight of his convoy, but be watchful to defend them from attack or surprise; and if distressed, to afford them all necessary assistance. He is to extend the same protection to his convoy when the United States are not engaged in war.

4. If the master of a ship shall misbehave, by delaying the convoy, abandoning, or disobeying the established instructions, the commander is to report him, with a narrative of the facts, to the secretary of the navy, by the first opportunity.

5. The commander is to carry a top-light in the night, to prevent separation, unless, on particular occasions, he may deem it improper.

6. He may order his signals to be repeated by as many ships of war under his command, as he may think fit.

7. When different convoys set sail at the same time, or join at sea, they are to keep together so long as their courses lie together: when it thus happens, the eldest commander of a convoy shall command in the first post; the next eldest, in the second; and so on according to seniority.

8. Commanders of different convoys are to wear the lights of their respective posts, and repeat the signals, in order, as is usual to flag-officers.

9. Convoys are to sail like divisions, and proper signals to be made at separation.

THE President of the United States of America ordains and directs the commanders of squadrons, and all captains and other officers in the navy of the United States to execute, and cause to be executed, the aforesaid regulations.

By command,

Secretary of the Navy.

WASHINGTON CITY: PRINTED FOR THE NAVY OFFICE. 1814.
[These regulations are a revised version of those first published by the Navy Department on 25 January 1802. The table of contents has not been reproduced here.]

U.S. Navy Uniform, 1802–13

These regulations remained in force until 1 January 1814 when they were superceded by a revised naval dress code.

THE *UNIFORM DRESS* OF THE CAPTAINS AND CERTAIN OTHER OFFICERS OF THE

Navy of the United States.

CAPTAINS' FULL DRESS.

THE COAT of blue cloth, with long lappels and lining of the same; a standing collar, and to be trimmed with gold lace, not exceeding one half inch in breadth, nor less than three eighths of an inch; in the following manner, to wit:—To commence from the upper part of the standing collar, and to descend round the lappels to the bottom of the coat; the upper part of the cuffs, round the pocket flaps and down the folds with

one single lace; four buttons on the cuffs and on the pocket flaps, nine on the lappels, and one on the standing collar; a gold epaulet on each shoulder; the buttons of yellow metal, with the foul anchor and American eagle, surrounded with fifteen stars; the button-holes to be worked with gold thread.

Vest and Breeches, white. The vest single breasted, with flaps and four buttons to the pockets, the buttons the same as the coat, only proportionably smaller.

THE UNDRESS—The same as the full dress, excepting the lace and the gold worked button-holes.

LIEUTENANTS' full dress.

The Coat of blue cloth, with long lappels, and lining of the same, with nine buttons on each lappel; a standing collar, and three buttons on the cuffs and on the pockets, the button-holes laced with such lace as is directed for the captain's; one epaulet on the left shoulder, except when acting as commanding officer, and then to be changed to the right shoulder.

Vest and Breeches—The same as the captain's, except three buttons and button-holes on the pockets of the vest.

THE UNDRESS—The same as the full dress, excepting the lace.

U.S. Navy, Full Dress, 1802

Company of Military Historians ®

MIDSHIPMENS' full dress.

The Coat of blue cloth, with lining and lappels of the same; the lappels to be short, with six buttons; standing collar, with a diamond formed of gold lace on each side, not exceeding two inches square; a slash sleeve, with three small buttons; all the button-holes to be worked with gold thread.

Vest and Breeches, white, the same as the lieutenants, except the buttons on the pockets of the vest.

THE UNDRESS—A short coat without worked button-holes, a standing collar with a button and a slip of lace on each side.

A Midshipman, when he acts as lieutenant, *by order of the Secretary of the Navy*, will assume the uniform of a lieutenant.

Captains and Lieutenants, when in full dress, to wear shoes, buckles, small swords, and gold laced cocked hats; the lace not to shew more than three quarters of an inch on each side—in undress to wear hangers.[1]

Midshipman, when in full dress, to wear gold laced cocked hats and hangers, with shoes and buckles.

Dirks, not to be worn on shore by any officer.

SURGEONS' full dress.

The Coat of blue cloth, with long lappels and lining of the same, nine navy buttons, with gold frogs[2] on the lappels, standing collar the same as the coat, and two gold frogs on each side of the collar, three navy buttons

below the pockets, and three gold frogs on the pocket flaps, and the same number of navy buttons to the cuffs, with gold frogs.

VEST AND BREECHES, white, with navy buttons—cocked hat.

SURGEONS' MATES' FULL DRESS.

THE COAT of blue cloth, with long lappels and lining of the same, nine navy buttons, and button-holes worked with gold thread; standing collar the same as the coat, with two navy buttons, and worked button-holes on each side; three navy buttons below the pockets, and three worked button-holes on the flaps—the same number of buttons on the cuffs, with worked button-holes.

VEST AND BREECHES, white, and cocked hat.

SAILING-MASTERS' FULL DRESS.

THE COAT of blue cloth; with standing collar, long lappels and lining of the same, nine buttons on the lappels and one on the standing collar, with a slip of lace; slash sleeves with three buttons, and three buttons to the pockets.

VEST AND BREECHES, white, and plain—cocked hat.

PURSERS' FULL DRESS.

THE COAT of blue cloth, with standing collar, long lappels and lining of the same, with nine buttons on the lappels, cuffs open behind, with three buttons, two above and one below; the collar to be embroidered with gold, not exceeding three-fourths of an inch in breadth, with one button on each side, the folds to have each three buttons, and three buttons under each pocket, the buttons the same as worn by the other officers of the navy.

VEST AND BREECHES, white—cocked hat.

A COMMODORE to have on each strap of his epaulets a silver star.

R. SMITH.

Navy Department, August 27, 1802.

Rare Book Room, Navy Department Library.

1. Hanger: A label commonly given to short, cut-and-thrust, fighting swords.

2. Frog: A fastener for garments made usually of braid in an ornamental loop design.

U.S. Navy Uniform, 1814–21

In response to the suggestions of a number of senior officers for a more comfortable uniform, Secretary Jones issued a revised naval dress code on 23 November 1813. The new uniform regulations took effect on 1 January 1814 and remained in force until 1 May 1821.

THE UNIFORM DRESS OF

THE OFFICERS OF THE NAVY

OF THE UNITED STATES.

CAPTAINS' FULL DRESS.

THE COAT of blue cloth, with broad lappels and lining of the same; a standing collar, and to be trimmed with gold lace, not exceeding one half inch in breadth, nor less than three eighths of an inch, in the following manner, to wit: Round the standing collar, and to descend round the lappels to the bottom of the coat; the upper part of the cuffs, round the pocket flaps, and down the folds, with one single lace, four buttons on the cuffs and on the pocket flaps, nine on the lappels, and one on the standing collar, two gold epaulets; the buttons of yellow metal, with the foul anchor and American eagle, surrounded with fifteen stars.

Pantaloons and Vest white. The vest single breasted, four buttons to the pockets, the buttons the same as the coat, only proportionably smaller.

UNDRESS.

The same as the full dress, excepting the lace and a rolling cape instead of standing collar.

MASTERS COMMANDANT.

Full dress and undress the same as a captain, excepting the Master Commandant to wear one epaulet on the right shoulder. No button on the cape or collar, and no lace round the pocket flaps.

LIEUTENANTS' FULL DRESS.

The Coat of blue cloth, with broad lappels and lining of the same, with nine buttons on each lappel; a standing collar with one button, and three buttons on the cuffs and on the pockets, laced with such lace as is directed for the captain's round the collar and cuffs; one epaulet on the left shoulder, except when acting as commanding officer, and then to be changed to the right shoulder.

Pantaloons and Vest the same as the captain's, except three buttons and button-holes on the pockets of the vest.

UNDRESS.

The same as the full dress, excepting the lace, and a rolling cape instead of standing collar.

MIDSHIPMEN'S FULL DRESS.

The Coat of blue cloth, with lining and lappels of the same; the lappels to be short, with six buttons; standing collar, with a diamond formed of gold lace on each side, not exceeding two inches square; with no buttons on the cuffs or pockets.

Pantaloons and Vest white, the same as the lieutenant's, except the buttons on the pockets of the vest.

UNDRESS.

A short coat, rolling cape, with a button on each side.

A Midshipman, when he acts as lieutenant, *by order of the Secretary of the Navy,* will assume the uniform of a lieutenant.

Captains and Lieutenants, when in full dress, to wear half boots, cut and thrust swords with yellow mountings, and gold laced cocked hats; the lace not to show more than three-quarters of an inch on each side.

Midshipmen, when in full dress, to wear plain cocked hats, half boots, and swords as above.

HOSPITAL SURGEONS,

Or NAVAL SURGEONS, *acting as such, by order of the Secretary of the Navy.*

A Coat of blue cloth, with broad lappels, and lining of the same, nine navy buttons on the lappels; standing collar the same as the coat, three navy buttons below the pockets, and the same number of buttons on the cuffs; *two rows* of gold lace, not exceeding one quarter of an inch broad, around the upper edge of the cuffs, and around the collar; one laced button hole on each side of the collar, with a navy button.

Pantaloons and Vest white, with navy buttons; plain cocked hat, half boots, and small sword.

UNDRESS.

The same as the full dress, excepting the lace on the cuffs, and instead of a standing collar, a rolling cape edged with gold cord.

SURGEONS' FULL DRESS.

The Coat of blue cloth, with broad lappels and lining of the same, nine navy buttons on the lappels, standing collar the same as the coat, and two laced button-holes on each side of the collar, three navy buttons below the pockets, and the same number of navy buttons to the cuffs.

Pantaloons and Vest white, with navy buttons; cocked hat, plain, half boots and small sword.

UNDRESS.

The same as the full dress, excepting the laced button holes, and rolling cape instead of a standing collar.

SURGEONS' MATES' FULL DRESS.

The Coat of the blue cloth, with broad lappels, and lining of the same, nine navy buttons; standing collar, the same as the coat, with one navy button and laced button-hole on each side; two navy buttons below the pockets, and the same number of buttons on the cuffs.

Pantaloons and Vest white, cocked hat, plain, half boots, and dirk.

UNDRESS.

Same as full dress, except rolling cape, with two buttons on each side without lace.

SAILING MASTERS' FULL DRESS

The Coat of blue cloth, with standing collar, broad lappels, and lining of the same, nine buttons on the lappels, and one on the standing collar, two buttons to the pockets, and the same number on cuffs.

Pantaloons and Vest white, and plain cocked hat, half boots, cut and thrust sword as before directed.

UNDRESS.

Same as full dress, except rolling cape, with one button without lace.

PURSERS' FULL DRESS.

The Coat of blue cloth, with standing collar, broad lappels and lining of the same, with nine buttons on the lappels, cuffs open behind with three buttons, two above and one below; the collar to be laced round with gold lace, not exceeding one half of an inch in breadth, with one button on each side, the folds to have each three buttons, and three buttons under each pocket, the buttons the same as worn by the other officers of the navy.

Pantaloons and Vest white; cocked hat, plain, half boots, and dirk.

UNDRESS.

Same as full dress, except rolling cape, with no button and no lace.

A COMMODORE to have on each strap of his epaulets a silver star.

MASTER'S MATES same as MIDSHIPMEN, except two buttons on the cuffs.

BOATSWAINS, GUNNERS, CARPENTERS, SAILMAKERS, short blue coats, with six buttons on the lappels, rolling cape, blue pantaloons, white vests, round hats, with cockade. No side-arms.

All other officers permitted to wear blue pantaloons, round hats, and dirks in undress.

Rare Book Room, Navy Department Library. (For an illustration of U.S. Navy officers and seamen in full dress for the period 1812–15, see p.11, A Ranked Society.)

Authorized Strength, 44-Gun Frigate

	1797	1807	1816
Captain	1	1	1
Lieutenants	4	5	6
Surgeon	1	1	1
Surgeon's Mates	2	2	2
Chaplain	1	1	1
Purser	1	1	1
Sailing Master	1	1	1
Midshipmen	8	16	20
Boatswain	1	1	1
Gunner	1	1	1
Sailmaker	1	1	1
Carpenter	1	1	1
Master's Mates	2	2	2
Captain's Clerk	1	1	1
Boatswain's Mates	2	2	2
Gunner's Mates	2		1
Sailmaker's Mates	1		1
Carpenter's Mates	2	2	1
Armorer	1	1	1
Cooper	1	1	1
Steward	1	1	1
Master-at-arms	1	1	1
Coxswain	1	1	1
[cont.]	1797	1807	1816
Boatswain's Yeoman			1
Gunner's Yeoman	1	1	1
Carpenter's Yeoman			1
Quarter Gunners	11	12	10
Quartermasters		12	8
Ship's Corporal			1
Cook	1	1	1
Able Seamen	150	120	150
Ordinary Seamen and Boys	103	172	170
First Lieutenant of Marines	1	1	1
Second Lieutenant of Marines	1	1	1
Sergeants	3	2	3
Corporals	3	2	2
Music (Fifers and Drummers)	2	2	2
Privates	50	50	48
Total	364	420	450
Total by rank			
Commissioned Officers	9	10	12
Warrant Officers	14	22	25
Petty Officers	28	38	36
Seamen	253	292	320
Marines	60	58	57

The figures for 1797 are drawn from "*An Act providing a Naval Armament*," in *Statutes at Large of the United States of America, 1789–1845*, 8 vols. (Boston: Little, Brown, 1846–67), 1:523–25. Those for 1807 and 1816 are taken from tables in *American State Papers*, 6, *Naval Affairs* (Washington, D.C.: Gales and Seaton, 1832), 1:172, 402.

Chronology of *Constitution* in the War of 1812

Dates in red denote *Constitution*'s major engagements.

1812

June

11 *Constitution* departs Washington Navy Yard, drops down the Potomac River, and anchors off the port town of Alexandria, Va. Over the next week, new hands are shipped, and the crew readies the frigate for sea, taking in stores for all departments, setting up the rigging, and making repairs.

13 At noon Secretary of the Navy Paul Hamilton, Washington Navy Yard Commandant Thomas Tingey, and Marine Corps Commandant Lieutenant Colonel Franklin Wharton visit *Constitution*.

18 Congress declares war against Great Britain. *Constitution* departs Alexandria and stands down the Potomac. Preparations for sea continue as the frigate works downriver. Shortly before midnight Hull receives sailing instructions from Secretary Hamilton enclosing a copy of the declaration of war. Hamilton orders Hull to complete *Constitution*'s crew and outfit at Annapolis, Md., and then sail for New York.

19 Regular drill at the great guns commences. Lieutenant George C. Read reads the declaration of war to the crew. On hearing this news, the men request permission to cheer, which is granted.

21 John Contee, 2d lieutenant of Marines, departs ship for Norfolk, Va., to receive *Constitution*'s Marine detachment.

23 The ship's remaining 24-pounder long guns are taken on board.

25 *Constitution* receives its battery of twenty-four 32-pounder carronades.

27 *Constitution* clears the mouth of the Potomac and stands up the Chesapeake Bay to Annapolis. At 5:30 p.m. the frigate enters Annapolis Roads.

28 *Constitution* moors off Annapolis. Readying the frigate for sea continues, including the shipping of new hands and the stationing of the crew.

July

3 Lieutenant Contee returns to the ship with its Marine detachment. Secretary Hamilton orders Hull to place himself under Commodore John Rodgers's command on reaching New York.

5 *Constitution* departs Annapolis and stands down Chesapeake Bay.

11 Final personnel come on board while anchored at the mouth of the Chesapeake.

12 *Constitution* weighs and stands to sea.

16 At 2:00 p.m. *Constitution*'s lookout spies four unknown sail. At 6:15 p.m. the frigate stands for the easternmost ship. At 7:30 p.m. Hull orders beat to quarters. At 11:15 p.m., its recognition signals ignored, *Constitution* turns away from the enemy vessel. The "Great Chase" is on. Over the next fifty-seven hours a six-vessel enemy squadron sails in pursuit of *Constitution*.

19 At 8:15 a.m. the British squadron gives over the chase.

20 Hull writes Secretary Hamilton that he is making for Boston.

26 *Constitution* enters Boston Bay and anchors off the lighthouse. Officers go ashore to procure provisions and open a naval rendezvous.

27 *Constitution* is towed to its anchorage in President Roads. Lighters begin provisioning and watering the ship. Captain Hull is greeted ashore with cheers as he passes up State Street.

28 Secretary Hamilton orders Hull to deliver up command of *Constitution* to William Bainbridge on his arrival in port, after which he is to proceed to Washington to assume command of *Constellation*.

29 Secretary Hamilton issues new instructions directing Hull to await further orders at Boston. These orders and those of the 28th arrive in Boston after *Constitution* has sailed.

August

2 *Constitution* sails from Boston steering a northeasterly course toward the coasts of Nova Scotia and Newfoundland.

10 *Constitution* captures and burns British merchant brig *Lady Warren*.

11 *Constitution* captures and burns British merchant brig *Adeona*.

15 *Constitution* captures *Adelaide* (*Adeline*). Midshipman John R. Madison and prize crew of five seamen are placed on board. *Adelaide* is recaptured by H.M. frigate *Statira* before making port in America.

19 *Constitution* captures H.M. frigate *Guerriere,* earning the nickname "Old Ironsides."

20 *Guerriere* set afire and blown up.

29 *Constitution* anchors in Nantasket Roads, Boston Harbor. Wounded prisoners are sent ashore to the hospital on Quarantine [Rainsford] Island.

31 *Constitution* anchors off Long Wharf. The prisoners are transferred to a prison ship. Over the next fifty-six days *Constitution* will undergo an extensive refit in preparation for its next cruise. John Rodgers's squadron (consisting of *President, United States, Congress, Hornet,* and *Argus*) also arrives in Boston this day.

Brig-rigged merchant vessels such as *Lady Warren*, *Constitution*'s first wartime prize, were captured in large numbers by American privateersmen and Navy warships. *Elements and Practice of Rigging and Seamanship, Vol. 1, 1794*

September

1 Hull receives a 17-gun salute from the Washington Artillery Company after he comes ashore at Long Wharf. *Constitution* returns the salute gun for gun. Crowds of Bostonians cheer Hull as he makes his way up State Street. Hull writes the Navy Department requesting a leave of absence to tend to family affairs.

2 Bainbridge writes to the Navy Department requesting command of *Constitution*.

3 Sixty-five of *Constitution*'s officers and men volunteer for temporary duty in *President* to sail in pursuit of a British squadron reported off Boston. They return to *Constitution* two days later.

5 Hull is honored with a public dinner at Faneuil Hall.

9 *Constitution* moors off the Navy Yard. Secretary Hamilton appoints William Bainbridge to command a three-ship squadron in *Constitution*. On sailing, Bainbridge's squadron is to annoy the enemy and protect American commerce. The other squadron vessels include the frigate *Essex,* David Porter, and the sloop of war *Hornet,* James Lawrence.

11 A cartel carrying Constitutions captured in the prize ship *Adelaide* arrives at Boston.

15 Bainbridge assumes command of *Constitution* from Hull. The crew voices its dissatisfaction with their new captain. One crewman is placed in confinement aboard Gunboat *No. 85* for "mutinous expressions." Two days later, another crewman is confined in Gunboat *No. 85* on the same charge.

October

2 Lieutenant George Parker of *United States* reports to *Constitution* as first lieutenant. Parker replaces Charles Morris who was severely wounded in the action with *Guerriere*. Hamilton informs Bainbridge he is at liberty to take his squadron to sea whenever he chooses.

13 Bainbridge issues cruising orders to David Porter who is preparing *Essex* for sea at the Philadelphia Navy Yard. Bainbridge's instructions provide places and dates where the three squadron vessels might unite, identifying their cruising grounds as the waters off the Brazilian coast.

16 *Constitution* drops down from the navy yard and anchors near Long Wharf.

21 *Constitution* shifts its anchorage to President Roads, Boston Harbor.

22 Crew begins daily exercise at quarters preparatory to sailing.

24 *Hornet* joins *Constitution* in President Roads.

27 *Constitution* and *Hornet* stand to sea.

28 *Essex* clears the Delaware Capes and stands to sea.

November

8 *Constitution* and *Hornet* overhaul the American brig *South Carolina*. The merchantman is sent in as a prize for carrying a British trading license. *Hornet*'s Lieutenant William S. Cox is given charge of the prize.

26 A court-martial is convened aboard *Constitution* to try one of the frigate's Marines, Private John Pershaw. The court, composed of officers from *Hornet* and *Constitution,* finds Pershaw guilty, sentencing him to receive fifty lashes.

December

2 *Constitution* and *Hornet* arrive at Fernando de Noronha, Brazil, an island about 250 miles northeast of Cape São Roque, Brazil. The American ships take on water and fresh provisions (fish, pork, eggs, nuts, and fruit).

4 The American squadron sails from Fernando de Noronha. Bainbridge leaves David Porter a coded message detailing his intended cruising grounds. *Essex* arrives at the island ten days later.

6 *Constitution* and *Hornet* arrive off the South American mainland. Over the next week, the American warships run southward down the coast of Brazil for the port city of Salvador (also known as Bahia).

12 At dinnertime, *Constitution*'s crew gathers on deck in a "mutinous manner" and complains to Bainbridge about their short rations of bread and water. The ship's company has been on reduced rations since leaving Boston. After hearing Bainbridge's explanation, the men return below decks as ordered.

13 Bainbridge orders Lawrence to enter port of Salvador to obtain provisions and to make contact with U.S. Consul Henry Hill. *Hornet* anchors in Salvador harbor the following day. A British 20-gun ship, *Bonne Citoyenne,* rumored to be carrying over a million dollars in gold specie, lies at anchor in the harbor.

18 *Hornet* departs Salvador and rejoins *Constitution* after procuring water, jerked beef, fruit, and oxen. Provisions are transferred to *Constitution* through the following day.

19 *Constitution* cruises to the north of Salvador in quest of enemy vessels while *Hornet* patrols the waters off the Brazilian city.

24 *Constitution* and *Hornet* reunite off Salvador.

26 *Constitution* stands to sea again leaving *Hornet* to blockade *Bonne Citoyenne*.

29 *Constitution* captures H.M. frigate *Java*.

31 *Java* is set afire and blown up. Secretary of the Navy Paul Hamilton resigns.

1813

January

1 *Constitution* anchors in Salvador harbor to land prisoners, take on provisions, and repair battle damage. Four days later, the frigate stands out of the harbor.

6 *Constitution* sails for home leaving *Hornet* on station off Salvador. On its return passage home, *Hornet* defeats and sinks H.M. brig-sloop *Peacock* off the coast of Guyana.

19 William Jones assumes secretaryship of the Navy Department.

26 After repeated failures to rendezvous with *Constitution* and *Hornet,* David Porter sails *Essex* into the Pacific where, after enjoying initial success against British whalers in the Galápagos, his ship is captured in Valparaíso, Chile, in February 1814.

29 Congress votes Hull a gold medal and his commissioned officers silver medals in recognition of their gallant conduct in the action with *Guerriere.*

February

15 *Constitution* arrives in Boston Bay. Boisterous weather and contrary winds prevent the frigate from entering the city's harbor.

18 Bainbridge comes ashore at Boston's Long Wharf. He is honored that night with a banquet at the Exchange Coffee House.

21 Bainbridge reports that *Constitution* requires an extensive overhaul including the replacement of beams, decking, waterways, ceiling, rigging, sails, and copper sheathing. Poor weather, lack of storage facilities, and a shortage of key timbers delay work on the frigate. All major repairs to the ship's structure, except coppering, are completed by the third week of June.

March

3 Congress awards $50,000 to Hull and his crew for the capture and destruction of *Guerriere.* A like sum is awarded to Bainbridge and his crew for the capture and destruction of *Java.* Congress also votes Bainbridge a gold medal and his commissioned officers silver medals in testimony of their brave and skilful combat with *Java.*

15 Secretary Jones directs Bainbridge to assume command of the Charlestown Navy Yard. He replaces Hull who has been appointed commandant of the Portsmouth Navy Yard.

April

18 Bainbridge transfers one hundred Constitutions to Isaac Chauncey's command on Lake Ontario. A similar transfer of fifty more crewmen takes place before the end of the month. The majority of the Old Ironsides's company has now left the ship through transfer or expiration of service.

May

4 Secretary Jones appoints recently promoted James Lawrence to command *Constitution.* Two days later, the Navy Secretary revokes the order, assigning Lawrence instead to the command of the frigate *Chesapeake* at Boston.

7 Jones orders Charles Stewart, captain of the frigate *Constellation* at Norfolk, to assume command of *Constitution.*

June

22 Charles Stewart arrives in Boston. His first priority is entering a new crew for *Constitution.* Slow recruiting and heavy desertion over the summer and fall months delay completing the ship's company until the second week of December.

September

19 Secretary Jones issues sailing orders to Stewart. He instructs the latter to sail to the coast of Cayenne (French Guyana), follow a northern trek through the Windward and Leeward Islands, and then choose between continuing operations in the West Indies or cruising in European waters. The destruction of enemy commerce is Stewart's main objective.

November

5 *Constitution* is ready for sea, save a shortage of crewmen. Among the innovations Stewart has introduced in fitting out the ship are a furnace

for heating 24-pound shot and two large tanks installed on the orlop deck for storing the ship's beef. This latter novelty results in a great spoilage of *Constitution*'s beef.

December

30 *Constitution* departs Boston Harbor steering a southeasterly course. The frigate is fully provisioned for a six-month cruise.

1814

January

9 The crew refuses to consent to a reduction in their bread and spirit ration for the purposes of extending the ship's cruise.

February

1 The majority of the ship's messes agree to a reduction in their beef and bread ration.

14 *Constitution* captures the British merchant ship *Lovely Ann* and H.M. schooner *Pictou*. After its cargo has been discharged into the sea, *Lovely Ann* is placed under Midshipman Pardon M. Whipple's charge and sails for Barbados with the British prisoners. *Pictou* is set afire.

16 *Lovely Ann* arrives at Barbados. British authorities refuse to recognize the *Lovely Ann* as a cartel and detain Whipple until August.

17 *Constitution* captures and scuttles the British merchant schooner *Phoenix*.

19 *Constitution* captures and scuttles the British merchant brig *Catherine*.

March

21 Several cases of scurvy are reported among the crew.

27 A defect in the ship's mainmast is observed.

April

3 *Constitution* is chased into Marblehead, Mass., by the British frigates *Junon* and *Tenedos*. To elude his pursuers, Stewart orders the ship lightened by pumping water and throwing provisions and stores overboard. Later that day, *Constitution* shifts its anchorage to Salem harbor.

4 Stewart sends his official report to Secretary Jones. In it, he cites a looming shortage of provisions, the appearance of scurvy among the crew, and the likelihood of avoiding blockaders off Boston as reasons for his early return.

8 Captain Stewart and *Constitution*'s officers attend a dinner and ball held in their honor at Salem.

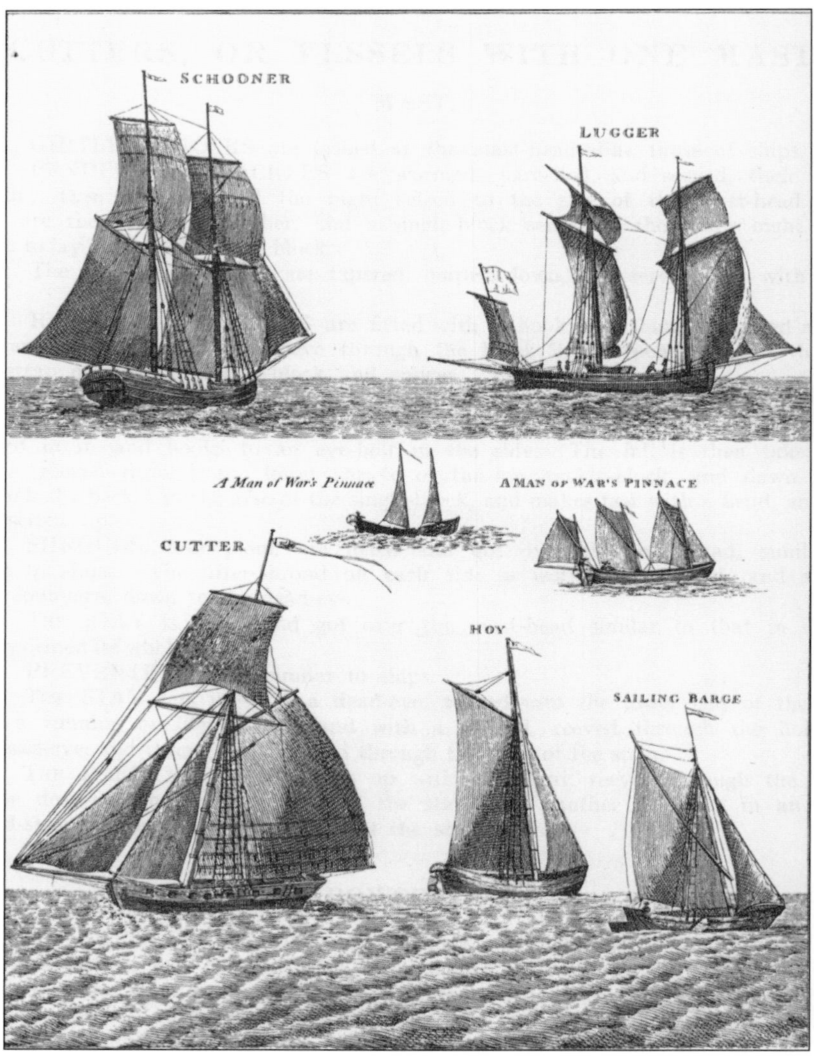

Phoenix, a schooner-rigged British merchant vessel, fell prey to *Constitution* on 17 February 1814. *Elements and Practice of Rigging and Seamanship, Vol. 1, 1794*

17 At noon *Constitution* departs Salem, anchoring at Boston seven hours later. The cheers of thousands gathered on the city's wharves greet the frigate.

20 Jones orders Bainbridge to oversee the outfitting of *Constitution* for another cruise.

21 Dissatisfied with Charles Stewart's reasons for returning to port, Secretary Jones orders Bainbridge to convene a court of inquiry to examine into the causes of *Constitution*'s prematurely terminated cruise. Captain Oliver H. Perry is appointed the other member of the court.

May

3 Bainbridge convenes Stewart's court of inquiry. In his testimony, Stewart cites damage to *Constitution*'s mainmast and the worn-out condition of its sails as additional reasons for his early return to port.

9 The court finds Stewart guilty of an error in judgment only. Based on the court's findings, Jones determines no further proceedings against Stewart are merited.

21 Jones issues new sailing orders to Stewart, directing him to take station off the Grand Banks to intercept British transports and supply ships bound for Canada, then to cruise northwest of the Azores to capture West Indian merchantmen bound for England. Enemy blockaders prevent *Constitution*'s departure until December. During the summer months the ship's crew will assist in the defense of the harbor, including the construction of batteries on Noddles Island.

November

29 Stewart is issued revised sailing instructions. He is to cruise east of Bermuda to intercept inbound enemy transports and store ships, then to shape a course for the Iberian Peninsula, ranging along the coasts of Spain and Portugal to capture vessels in the trek of the West Indian trade.

December

1 William Jones departs office as Navy secretary.

17 *Constitution* sails from Boston.

19 Benjamin W. Crowninshield of Massachusetts is appointed secretary of the navy.

24 *Constitution* captures the British merchant brig *Lord Nelson*. American and British diplomats sign a peace treaty at Ghent, Belgium.

25 *Lord Nelson* is scuttled.

1815

January

16 Benjamin W. Crowninshield assumes secretaryship of the Navy Department.

February

8 *Constitution* obtains intelligence of the Anglo-American peace treaty from two vessels out of Ireland.

15 The Treaty of Ghent reaches the United States.

16 *Constitution* captures the British merchant ship *Susannah*.

17 Congress ratifies the Treaty of Ghent. All vessels taken by belligerents within twelve to ninety days of the treaty's ratification (depending on which part of the world the capture was made) are considered legal prizes.

19 *Susannah* is sent in to the United States as a prize.

20 *Constitution* captures H.M. ships *Cyane* and *Levant* seventy miles east-northeast of Madeira. Lieutenant Beekman V. Hoffman and a small prize crew are placed aboard *Cyane*.

21 *Constitution*'s first lieutenant, Henry E. Ballard is given charge of *Levant*. Prisoners and baggage are transferred to *Constitution*. Repairs to prizes underway.

March

1 *Constitution* and its prizes steer for the Cape Verde Islands.

8 *Constitution* and its prizes anchor in roads of Maio, one of the Cape Verde Islands.

9 The frigate takes on provisions including a bullock for the sick and wounded. Later that day, the three ships get underway and stand for São Tiago.

10 *Constitution* and its prizes anchor in Port Praia, São Tiago, Cape Verde Islands.

11 When three enemy sail (the frigates *Acasta, Leander,* and *Newcastle*) are spied standing for Port Praia, Stewart gives the signal to get under way. *Cyane* and *Constitution* escape. Unable to outsail its pursuers, *Levant* attempts to regain the anchorage in Port Praia but is recaptured. *Constitution* stands to the westward in quest of vessels bound to England from points south of the equator.

15 With *Constitution* now shorthanded, Stewart reorganizes the frigate's quarter bill placing Marines at guns and giving command of the quarterdeck guns to Marine officers and Purser Robert Pottinger.

April

2 *Constitution* arrives at São Luís on the island of Maranhão, Brazil.

4 British prisoners are discharged ashore.

6 The frigate begins taking on food and water over the next week.

9 *Cyane* anchors in the lower end of the Hudson River (North River), New York.

15 *Constitution* puts to sea again after fighting contrary winds for two days to exit São Luís.

28 *Constitution* arrives off San Juan, Puerto Rico. Stewart sends a boat ashore to "obtain information relative to our national affairs." Lieutenant William M. Hunter returns with news that a peace treaty has been signed and ratified by Congress. The frigate stands to the northward the following day.

May

2 Lieutenant Ballard reports his arrival at Baltimore along with part of the prize crew captured in *Levant*.

15 *Constitution* anchors off Sandy Hook. Stewart dispatches his official report to the Navy Department.

16 *Constitution* anchors off the Battery at New York, firing a 15-gun salute.

18 Secretary of the Navy Benjamin W. Crowninshield orders *Constitution* to Boston.

25 Secretary Crowninshield orders Stewart to supply enough men, by draft or volunteer, to man the frigate *Congress*.

27 *Constitution* arrives in outer harbor of Boston.

29 *Constitution* anchors off the town. Gun salutes are fired to honor the frigate. Stewart is escorted to the Exchange Coffee House.

June

16 Stewart requests a furlough. Secretary Crowninshield grants his request on the 25th.

July

16 Stewart departs *Constitution* on furlough.

1816

February

22 Congress votes Stewart a gold medal and his commissioned officers silver medals in recognition of their gallant conduct in the action with *Cyane* and *Levant*.

April

26 Congress awards Stewart and the crew of *Constitution* $25,000 for the capture of *Levant*.

List of *Constitution*'s Captures, 1812–15

Date	Name	Vessel Type	Commander	Crew	Bound From	Bound For	Cargo	Disposition
08/10/12	*Lady Warren*	Brig		7	St. John's, Newfoundland	Cape Breton	in ballast	Burnt
08/11/12	*Adeona*	Brig		7	Shediac, New Bruns-wick	Newcastle, upon Tyne	squared pined timber	Burnt
08/15/12	*Adelaide (Adeline)*	Brig		6	Bath, Maine	London	hard and dry goods	Recaptured
08/19/12	*Guerriere*	Royal Navy frigate	James R. Dacres	311				Burnt 08/20/12
11/08/12	*South Carolina*	American brig	Richard Gaul	9	Lisbon	Philadelphia	in ballast	Sent in as prize; restored to owners
12/29/12	*Java*	Royal Navy frigate	Henry Lambert	400				Burnt
02/14/14	*Lovely Ann*	Ship		16	Bermuda	Surinam	lumber, fish, and flour	Sent in as cartel to Barbados
02/14/14	*Pictou*	Royal Navy schooner	Edward L. Stephens	57				Burnt
02/17/14	*Phoenix*	Schooner	Captain Tynes		Demerara	Barbados	lumber	Scuttled
02/19/14	*Catherine*	Brig	Captain Smith		Grenada	St. Thomas	in ballast	Scuttled
12/24/14	*Lord Nelson*	Brig	William Hatchard	7	St. John's, Newfoundland	Bermuda	dry goods, slops, pork, gin, wine, etc.	Scuttled 12/25/14
02/16/15	*Susannah*	Ship	Malcolm Ross	16	Buenos Aires	Liverpool	tallow and hides	Sent in as prize
02/20/15	*Cyane*	Royal Navy ship	Gordon T. Falcon	180				Sent in as prize
02/20/15	*Levant*	Royal Navy ship	George Douglas	156				Recaptured 03/11/15

Burnt or scuttled	8	
Sent in as cartel	1	
Sent in as prizes	5	(two of which were recaptured and one which was restored to its owners)
Totals	14	

Figures appearing in the column marked "Crew" are taken from American sources. In the case of *Java*, *Lovely Ann*, *Pictou*, *Cyane*, and *Levant*, Bainbridge's and Stewart's correspondence supplied the data. Totals for *South Carolina*, *Lord Nelson*, and *Susannah* are derived from American admiralty court records. The remaining figures are drawn from *Constitution*'s logbooks.

Constitution's Commanders: The Operational Years

Samuel Nicholson, Sr.	22 July 1798 - 5 June 1799
Silas Talbot	5 June 1799 - 8 September 1801
Edward Preble	14 May 1803 - 28 October 1804
Stephen Decatur	28 October 1804 - 9 November 1804
John Rodgers, Sr.	9 November 1804 - 30 May 1806
Hugh G. Campbell	30 May 1806 - 8 December 1807
John Rodgers, Sr.	20 February 1809 - 17 June 1810
Isaac Hull	17 June 1810 - 15 September 1812
William Bainbridge	15 September 1813 - 18 July 1813
Charles Stewart	18 July 1813 - 16 July 1815
Jacob Jones	1 April 1821 - 31 May 1824
Thomas Macdonough	31 May 1824 - 14 October 1825
Daniel T. Patterson	14 October 1825 - 5 December 1825
George C. Read	23 January 1826 - 21 February 1826
Daniel T. Patterson	21 February 1826 - 19 July 1828
Jesse D. Elliott	3 March 1835 - 18 August 1838
Daniel Turner	1 March 1839 - 8 November 1841
Foxhall A. Parker, Sr.	15 July 1842 - 16 February 1843
John Percival	13 December 1843 - 5 October 1846
John Gwinn	9 October 1848 - 4 September 1849
Thomas A. Conover	18 September 1849 - 16 January 1851
John S. Rudd	22 December 1852 - 15 June 1855

U.S. Navy Ration, 1801–42

View of the Navy Ration, as established by act of March 3d, 1801, valued at 20 cents.												
	Bread	Flour	Beef	Pork	Suet	Cheese	Peas	Rice	Butter	Spirits	Molasses	Vinegar
	oz.	lb.	lb.	lb.	lb.	oz.	pint.	pint.	oz.	pt	pt	pt
Sunday	14	½	1¼	-	¼	-	-	-	-	½	-	-
Monday	14	-	-	1	-	-	½	-	-	½	-	-
Tuesday	14	-	1	-	-	2	-	-	-	½	-	-
Wednesday	14	-	-	1	-	-	-	½	-	½	-	-
Thursday	14	½	1¼	-	¼	-	-	-	-	½	-	-
Friday	14	-	-	-	-	4	-	½	2	½	½	-
Saturday	14	-	-	1	-	-	½	-	-	½	-	½

American State Papers, 6, Naval Affairs, (Washington, D.C.: Gales and Seaton, 1832), 1:183.

Constitution at a Glance

Constructor	George Claghorne
Designer	Joshua Humphreys
Built at	Hartt Shipyard, Boston, Mass.
Launched	21 October 1797
First sailed	22 July 1798
Crew(1812-era)	440–485
Length	
billethead to taffrail	204'
at waterline	175'
Width (beam)	43' 6"
Mast heights (keel to truck)	
fore	198'
main	220'
mizzen	172' 6"
Depth of hold	14' 3"
Draft	
forward	21' (approx.)
aft	23' (approx.)
Hull displacement	1,576 tons
Sails (1812-era)	up to 40 types used; nearly an acre of canvas under full-sail
Speed (under full sail)	13+ knots

Constitution's Armament

Commander	12-Pounder Long Guns	18-Pounder Long Guns	24-Pounder Long Guns	32-Pounder Carronades	24-Pounder Gunnades	Total Guns
Samuel Nicholson (1798)	14	16	30			60
Isaac Hull (1812)		1	30	24		55
William Bainbridge (1812)		1	30	24		55
Charles Stewart (1814)			30	22		52
Charles Stewart (1815)			30	20	2	52

...so well directed was the fire of the Constitution, and so closely kept up, that in less than thirty minutes, from the time we got alongside of the Enemy (One of their finest Frigates) she was left without a Spar Standing, and the Hull cut to pieces in such a manner as to make it difficult to keep her above water...

Bibliography & Index

Selected Bibliography

1. Narrative Histories of U.S. Frigate *Constitution*

Desy, Margherita M. "*Constitution*: Where Was She at 100?" *Nautical Research Journal* 42 (September 1997): 145–53.

Feuer, A. B. "First Frigate Duel of 1812." *Military History* 13 (February 1997): 30–36.

Gillmer, Thomas C. *Old Ironsides: The Rise, Decline, and Resurrection of the USS Constitution*. Camden, ME: International Marine, 1993.

Jennings, John. *Tattered Ensign*. New York: Thomas Y. Crowell Company, 1966.

Martin, Tyrone G. "Isaac Hull's Victory Revisited." *American Neptune* 47 (Winter 1987): 14–21.

______. *A Most Fortunate Ship: A Narrative History of Old Ironsides*. Rev. ed. Annapolis, MD: Naval Institute Press, 1997.

______. *Undefeated: "Old Ironsides" in the War of 1812*. Chapel Hill, NC: Tryon Publishing Co., 1996.

Rand, Anne Grimes. "'Old Ironsides' in War and Peace." *New England Journal of History* 53 (Spring 1996): 15–31.

2. Histories of U.S. Frigate *Constitution*'s Era

Fowler, William M. *Jack Tars and Commodores: The American Navy, 1783–1815*. Boston: Houghton Mifflin, 1984.

Gruppe, Henry E. *The Frigates*. Alexandria, VA: Time-Life Books, 1979.

Hickey, Donald R. *The War of 1812: A Forgotten Conflict*. Urbana: University of Illinois Press, 1989.

Langley, Harold D. *A History of Medicine in the Early U.S. Navy*. Baltimore, MD: The Johns Hopkins University Press, 1995.

Leiner, Frederick C. *The End of Barbary Terror: America's 1815 War Against the Pirates of North Africa*. New York: Oxford University Press, 2006.

______. *Millions for Defense: The Subscription Warships of 1798*. Annapolis, MD: Naval Institute Press, 2000.

McKee, Christopher. *A Gentlemanly and Honorable Profession: The Creation of the U.S. Naval Officer Corps, 1794–1815*. Annapolis, MD: Naval Institute Press, 1991.

Palmer, Michael A. *Stoddert's War: Naval Operations During the Quasi-War with France, 1798–1801*. Columbia: University of South Carolina Press, 1987.

Robotti, Frances D., and James Vescovi. *The USS Essex and the Birth of the American Navy*. Holbrook, MA: Adams Media Corporation, 1999.

Roosevelt, Theodore. *The Naval War of 1812*. 1882. Reprinted with an introduction by Edward K. Eckert. Annapolis, MD: Naval Institute Press, 1987.

Symonds, Craig L. *Navalists and Antinavalists: The Naval Policy Debate in the United States, 1785–1827*. Newark: University of Delaware Press, 1980.

Toll, Ian W. *Six Frigates: The Epic History of the Founding of the U.S. Navy*. New York: W. W. Norton & Company, 2006.

Tucker, Spencer C. *The Jeffersonian Gunboat Navy*. Columbia: University of South Carolina Press, 1993.

Wheelan, Joseph. *Jefferson's War: America's First War on Terror, 1801–1805*. New York: Carroll & Graf, 2003.

3. Biographies of U.S. Frigate *Constitution*'s Commanding Officers

Allison, Robert J. *Stephen Decatur: American Naval Hero, 1779–1820*. Amherst: University of Massachusetts Press, 2005.

Berube, Claude G., and John A. Rodgaard. *A Call to the Sea: Captain Charles Stewart of the USS Constitution*. Washington, DC: Potomac Books, 2005.

Biographical Sketch, and Services of Commodore Charles Stewart, of the Navy of the United States. Philadelphia: J. Harding, 1838.

Dearborn, Henry A. S. *The Life of William Bainbridge, Esq. of the United States Navy*. Edited by James Barnes. Princeton: Princeton University Press, 1931.

Ellis, James H. *Mad Jack Percival: Legend of the Old Navy*. Annapolis, MD: Naval Institute Press, 2002.

Fowler, William M., Jr. *Silas Talbot: Captain of Old Ironsides*. Mystic, CT: Mystic Seaport Museum, 1995.

Long, David F. *"Mad Jack:" The Biography of Captain John Percival, USN, 1779–1862*. Westport, CT: Greenwood Press, 1993.

______. *Ready to Hazard: A Biography of Commodore William Bainbridge, 1774–1833*. Hanover, NH: Published for University Press of New Hampshire by University Press of New England, 1981.

Maloney, Linda M. *The Captain from Connecticut: The Life and Naval Times of Isaac Hull*. Boston: Northeastern University Press, 1986.

McKee, Christopher. *Edward Preble: A Naval Biography, 1761–1807.* Annapolis, MD: Naval Institute Press, 1972.

Schroeder, John H. *Commodore John Rodgers: Paragon of the Early American Navy.* Gainesville: University Press of Florida, 2006.

Skaggs, David C. *Thomas Macdonough: Master of Command in the Early U.S. Navy.* Annapolis, MD: Naval Institute Press, 2003.

Tucker, Spencer. *Stephen Decatur: A Life Most Bold and Daring.* Annapolis, MD: Naval Institute Press, 2005.

4. Historical Documents and Eyewitness Reports Regarding U.S. Frigate *Constitution*

Dudley, William S., and Michael J. Crawford, eds. *The Naval War of 1812: A Documentary History.* 3 vols. to date. Washington, DC: Naval Historical Center, 1985– .

Evans, Amos A. *Journal Kept on Board the Frigate "Constitution," 1812.* N.d. Reprint, n.p.: Paul Clayton, 1928.

Fischer, Frederick C. *Experienced and Conquered: Frederick C. Fischer, Musician, U.S. Navy Aboard USS Constitution 1844–1846.* Edited by Annabelle F. Fischer. Translated by Noah G. Good. Westminster, MD: Peach Originals, 1996.

Hawes, Lilla M., ed. "Letters of Henry Gilliam, 1809–1817." *Georgia Historical Quarterly* 38 (March 1954): 46–66.

Humphreys, Assheton. *The USS Constitution's Finest Fight, 1815: The Journal of Acting Chaplain Assheton Humphreys, US Navy.* Edited by Tyrone G. Martin. Mount Pleasant, S.C.: Nautical & Aviation Publishing Company of America, 2000.

[Jones, George.] *Sketches of Naval Life, with Notices of Men, Manners and Scenery, on the Shores of the Mediterranean, in a Series of Letters from the Brandywine and Constitution Frigates.* 2 vols. New Haven, CN: Hezekiah Howe, 1829.

[Mercier, Henry J.] *Life in a Man-of-War: Scenes in "Old Ironsides" during Her Cruise in the Pacific.* Philadelphia: Lydia R. Bailey, 1841.

Morris, Charles. *The Autobiography of Commodore Charles Morris, U.S. Navy.* 1880. Reprinted with an introduction and additional notes by Frederick C. Leiner. Annapolis, MD: Naval Institute Press, 2002.

Smith, Moses. *Naval Scenes in the Last War; Or, Three Years on Board the Frigate Constitution, and the Adams; Including the Capture of the Guerriere.* Boston: Gleason's Publishing Hall, 1846.

Thomas, Henry G. *Around the World in Old Ironsides: The Voyage of USS Constitution, 1844–1846.* Edited by Alan B. Flanders. Lively, VA: Brandylane Publishers, 1993.

Whipple, Pardon M. *Letters from Old Ironsides, 1813–1815.* Edited by Norma A. Price. Tempe, AZ: Beverly-Merriam Press, 1984.

5. Construction, Repairs, and Model-making of U.S. Frigate *Constitution*

Bass, Ethyl L., and William P. Bass. *Constitution Second Phase 1802–07. . . Mediterranean, Tripoli, Malta & More.* Melbourne, FL: Shipsresearch, 1981.

Gilkerson, William. "Wind-borne Once More: U.S.S. *Constitution* Sails Again." *WoodenBoat* 139 (November/December 1997): 48-59.

Marden, Luis. "Restoring Old Ironsides." *National Geographic* 191 (June 1997): 38-53.

Marquardt, Karl H. *The 44-Gun Frigate USS Constitution, "Old Ironsides."* Anatomy of the Ship. Annapolis, MD: Naval Institute Press, 2005.

Martin, Tyrone G. *Creating a Legend.* Chapel Hill, NC: Tryon Publishing Co., 1997.

______. "Humphrey's Real Innovation." *Naval History* 8 (March/April 1994): 32–37.

______. "The USS *Constitution*: A Design Confirmed." *American Neptune* 57 (Summer 1997): 257–65.

Morrison, Christopher. "Technical Aspects of Preparing 'Old Ironsides' to Sail Again." *Nautical Research Journal* 42 (September 1997): 154–61.

Reilly, John C. *The Constitution Gun Deck.* Washington: Naval Historical Center, 1983.

Weitzman, David. *Old Ironsides: Americans Build a Fighting Ship.* Boston: Houghton Mifflin, 1997.

6. Related Web Sites

U.S. Department of the Interior. National Park Service. www.nps.gov/archive/bost/bost_lographics/cnyintro.htm

U.S. Navy. Naval Historical Center Detachment, Boston. www.history.navy.mil/constitution/index.html

USS *Constitution.* www.ussconstitution.navy.mil/

USS Constitution Museum. www.ussconstitutionmuseum.org/index.htm

Index